Roman von Wartburg

Eye Movements and Scene Perception

Roman von Wartburg

Eye Movements and Scene Perception

An Investigation of Basic Effects

Südwestdeutscher Verlag für Hochschulschriften

Impressum/Imprint (nur für Deutschland/ only for Germany)
Bibliografische Information der Deutschen Nationalbibliothek: Die Deutsche Nationalbibliothek
verzeichnet diese Publikation in der Deutschen Nationalbibliografie; detaillierte bibliografische
Daten sind im Internet über http://dnb.d-nb.de abrufbar.

Alle in diesem Buch genannten Marken und Produktnamen unterliegen warenzeichen-, marken-
oder patentrechtlichem Schutz bzw. sind Warenzeichen oder eingetragene Warenzeichen der
jeweiligen Inhaber. Die Wiedergabe von Marken, Produktnamen, Gebrauchsnamen,
Handelsnamen, Warenbezeichnungen u.s.w. in diesem Werk berechtigt auch ohne besondere
Kennzeichnung nicht zu der Annahme, dass solche Namen im Sinne der Warenzeichen- und
Markenschutzgesetzgebung als frei zu betrachten wären und daher von jedermann benutzt
werden dürften.

Verlag: Südwestdeutscher Verlag für Hochschulschriften Aktiengesellschaft & Co. KG
Dudweiler Landstr. 99, 66123 Saarbrücken, Deutschland
Telefon +49 681 37 20 271-1, Telefax +49 681 37 20 271-0, Email: info@svh-verlag.de
Zugl.: Diss. Univ. Bern, 2006

Herstellung in Deutschland:
Schaltungsdienst Lange o.H.G., Berlin
Books on Demand GmbH, Norderstedt
Reha GmbH, Saarbrücken
Amazon Distribution GmbH, Leipzig
ISBN: 978-3-8381-0592-5

Imprint (only for USA, GB)
Bibliographic information published by the Deutsche Nationalbibliothek: The Deutsche
Nationalbibliothek lists this publication in the Deutsche Nationalbibliografie; detailed
bibliographic data are available in the Internet at http://dnb.d-nb.de.
Any brand names and product names mentioned in this book are subject to trademark, brand
or patent protection and are trademarks or registered trademarks of their respective holders.
The use of brand names, product names, common names, trade names, product descriptions
etc. even without a particular marking in this works is in no way to be construed to mean that
such names may be regarded as unrestricted in respect of trademark and brand protection
legislation and could thus be used by anyone.

Publisher:
Südwestdeutscher Verlag für Hochschulschriften Aktiengesellschaft & Co. KG
Dudweiler Landstr. 99, 66123 Saarbrücken, Germany
Phone +49 681 37 20 271-1, Fax +49 681 37 20 271-0, Email: info@svh-verlag.de

Copyright © 2009 by the author and Südwestdeutscher Verlag für Hochschulschriften
Aktiengesellschaft & Co. KG and licensors
All rights reserved. Saarbrücken 2009

Printed in the U.S.A.
Printed in the U.K. by (see last page)
ISBN: 978-3-8381-0592-5

Abstract

In this thesis, my basic research on eye movements during scene perception is reported. The specific aspects under scrutiny arose from the insight that there is a considerable number of elementary factors whose influence on eye movements have been largely ignored in the studies to date. The present work indicates that many basic parameters are more important than previously thought, and pertain to theoretical as well as methodological issues.

In Experiment 1, the effect of image size on saccade amplitudes was investigated. In a recognition task, 48 participants viewed 96 colour scene images scaled to four different sizes, while their eye movements were recorded. The analyses revealed that mean and median saccade amplitude are directly proportional to image size, while the peak of the distribution lies at the lower end of the range of detectable saccades. However, image size does not significantly change the course of saccade amplitudes over the total viewing time. Irrespective of image size, saccade amplitudes initially increase and then gradually decrease. Moreover, a meta analysis was performed in which mean saccade amplitudes reported in earlier image perception studies were summarised. Overall, the results strongly suggest that, as far as mean and median saccade amplitudes are concerned, the size of stimulus images is the dominant factor. Other factors such as image type and content, viewing task, or measurement equipment, only are of subordinate importance.

The aim of Experiment 2 was to assess the influence of viewing task and stimulus repetition on basic oculomotor measures and the spatial pattern of fixations. Fifty-four participants were assigned to three groups which received different viewing instructions, free viewing, preparation for a recognition task, or a detail memory task. Forty-eight scene images were presented twice, with a delay of 20 minutes in-between. Of these three tasks, the recognition task caused longer fixation durations, a higher similarity of spatial fixation patterns between the first and second viewing, and a higher between-participant similarity of fixation patterns. This indicates that the viewing strategies adopted by the participants play a crucial role, as these results cannot be interpreted based on memory load effects alone. Regarding the effects of stimulus repetition, longer fixation durations and shorter saccades during the second presentation were observed. Moreover, results from both experiments underline that several aspects of oculomotor behaviour are not constant over the time a scene image is examined; the behaviour seems to be of a different quality during the first one or two seconds.

Acknowledgments

My special thanks go to Prof. Dr. René M. Müri, head of the Perception and Eye Movement Laboratory (Department of Neurology and Department of Clinical Research, University Hospital of Bern) for his valuable technical and personal support throughout the years I have been able to work in his lab. Then, I would like to thank Prof. Dr. Rudolf Groner and his team for feedback on my experimental work and advice in statistical matters at various stages of the project. Furthermore, my thanks go to the other researchers in our lab, lic. phil. Tobias Pflugshaupt, lic. phil. Pascal Wurtz, lic. phil. Mathias Lüthi, and Dr. Thomas Nyffeler, whose comments on experimental design issues were very helpful; they also reviewed earlier drafts of various subparts of this thesis. Moreover, I am thankful of the work Miriam Loertscher had been doing during her traineeship in our lab; she was responsible for stimulus preparation and data acquisition for the second experiment reported in this work. I will not forget to thank those people who indirectly supported this project by providing the laboratory infrastructure, Prof. Dr. Christian W. Hess (Department of Neurology) and Prof. Dr. André Haeberli (Department of Clinical Research). And last but not least my special thanks go to all participants who volunteered their time and thus made this research possible.

Table of contents

1 **Introduction** 1

2 **General theoretical part** 4
  2.1 Basic components of visual exploration behaviour 4
    2.1.1 Eye movements 4
      2.1.1.1 Retinal anatomy 4
      2.1.1.2 Saccades and fixations 7
      2.1.1.3 Other types of eye movements 8
    2.1.2 Eye movement control models 10
    2.1.3 Saccades 10
    2.1.4 Fixations 14
  2.2 Visual attention 19
    2.2.1 General considerations 19
    2.2.2 A two-component model of visual attention 21
  2.3 Visual attention and eye movements 27
  2.4 Spatial fixation patterns during scene perception 30
    2.4.1 Aspects of exploratory behaviour 30
    2.4.2 What guides visual exploration in scene perception? 32
      2.4.2.1 Bottom-up saliency 32
      2.4.2.2 Image content and informativeness 33
      2.4.2.3 Scene schemas 34
      2.4.2.4 Viewing task 35
    2.4.3 Summary 35

3 **Experiment 1: Image size and saccade amplitude** 38
  3.1 Theoretical part 38
  3.2 Methods 43
    3.2.1 Participants 43
    3.2.2 Stimuli 43
    3.2.3 Apparatus 44
    3.2.4 Experimental procedure 45
    3.2.5 Data analysis 47
  3.3 Results 49
  3.4 Meta analysis of earlier studies 58
  3.5 Discussion 65
    3.5.1 Overall saccade amplitude distribution 65
    3.5.2 The temporal course of saccade amplitude change 67

4 Experiment 2: Viewing task and stimulus repetition 69

4.1 Theoretical part 69
4.1.1 Viewing task 69
4.1.2 Repetition 70

4.2 Methods 72
4.2.1 Participants 72
4.2.2 Stimuli 72
4.2.3 Apparatus 73
4.2.4 Procedure 73
4.2.5 Data analysis 74
4.2.5.1 Basic oculomotor measures 75
4.2.5.2 Re-fixations 75
4.2.5.3 Scanpaths 75
4.2.5.4 Similarity index 77
4.2.5.5 Data visualisation 78

4.3 Results 79
4.3.1 Fixation duration 79
4.3.2 Saccade amplitude 82
4.3.3 Re-fixations 84
4.3.4 Scanpaths 85
4.3.5 Similarity index 86

4.4 Discussion 90
4.4.1 Viewing task 90
4.4.2 Stimulus repetition and temporal course 93

5 Summary and Conclusions 96
5.1 Theoretical aspects 96
5.2 Methodological aspects 98
5.3 Outlook 100

Name index *101*
Subject index *102*
References *105*
Appendix A: Images used in Experiment 1 .... *111*
Appendix B: Result tables for Experiment 1 .... *115*
Appendix C: Images used in Experiment 2 .... *117*
Appendix D: Result tables for Experiment 2 .... *119*

1 Introduction

Looking behaviour of humans in everyday situations is an enormously complex and multi-faceted phenomenon. In daily life, our visual system hast to deal with an enormous range of visual situations. In a single glance, we receive visual information from a range of almost 180 degrees in width and 120 degrees in height, and we are directly and indirectly able to discern visual depth information. Usually, the scenery is in full-colour and covers a wide range of luminance differences. The overall brightness affecting our eyes can vary between about 20,000 lux at noon on a clear summer's day and just a fraction of a lux when sitting in a dark room by the light of a single candle, but our visual system still shows an astonishing performance, outperforming the best cameras available today. Typically, natural scenes are not static, due to head and body movements of the spectator that cause a continuous change of perspective on the one hand, and due to the movements of animate beings and moving objects like vehicles, clouds, or trees swinging in the wind on the other hand. Most of the time, we are completely unaware of how we move our eyes, nor do we have to make a conscious effort to do so efficiently. Nevertheless, our looking behaviour is usually well tuned to the requirements of the visual situation as well as to our current goals and motivations.

Since the beginning of the instrumental investigation of human vision in the first half of the 19th century, many aspects of visual behaviour and performance have been studied, ranging from Weber's early psychophysical experiments to intricate experimental designs, combining cognitive measures with eye movement recordings which are possible with current technology. Over the last years, the central interest of my research has been to further elucidate the role of eye movements in daily visual behaviour. The focus of the current work lies on the question of how people look at scene images, in terms of directly observable aspects like *how* exactly the eyes move, *where* and *how long* people look, and trying to make some inferences about *why* people look at images in the way they do.

There are at least three good reasons to investigate eye movements. First and foremost, eye movements are interesting in their own right, and many conceivable effects of stimulus features on basic oculomotor measures have not been studied in depth. Previous research indicates that certain oculomotor measures – like saccadic speed for instance – are relatively constant across individuals and conditions, while others systematically vary with stimulus conditions as well as with the participants' internal states, like motivation or alertness.

— Introduction

Second, as far as perception of more complex stimuli is concerned, most studies of the performance in scene perception tasks have only used static, briefly presented images. However, vision is a dynamic process in which representations are built up over time from multiple eye fixations, ant thus eye movements are critical for the efficient acquisition of visual information during complex visuo-cognitive tasks. Therefore, the way in which the scanning process is controlled to service information acquisition is a central issue with high explanatory powers. Third, it is generally agreed nowadays that, under normal circumstances, eye movements are tightly coupled to visual attention. Therefore, eye movement recording is a valuable, non-invasive means for studying the temporal and spatial deployment of visual attention in any situation, and thus to gain insight into visuo-cognitive information processing.

Presumably, the task best investigated by means of oculographic methods is reading, because a text display is clearly structured and can be evaluated in a straightforward manner, followed by search tasks. For a long time, the sparse studies using more complex image material were descriptive in nature (e.g. Brandt, 1945; Buswell, 1935; Yarbus, 1967). The main reason for this is the fact that it is notoriously difficult to describe complex stimulus dimensions like semantic content, informativeness, saliency etc. in an objective way, in order to relate them to oculomotor behaviour. Therefore most researchers confined themselves to relatively simple stimuli like arrays of basic geometrical shapes or simple objects, line drawings, achromatic images and the like. Undoubtedly, experiments with simplified stimuli are indispensable, but a complete theory of natural vision requires experiments with ecologically more valid tasks and stimuli, such as viewing of everyday scenes. Nowadays, several eye tracking systems are commercially available which allow for eye movement recording with freely moving participants in almost any environment. However, measurement precision is restricted, and data evaluation is difficult as it is usually unavoidable to fall back on behavioural observation methods. With current technology, the maximum complexity that can be attained with high precision and reliability is presenting colour photographs of natural scenes or similarly complex images on a computer display and record the eye-movements simultaneously; complete head fixation is no longer required. In such set-ups, eye movements can unambiguously be related to image content.

As a consequence of technical progress, eye tracking equipment has become much more sophisticated, versatile, easy to use, and affordable in recent times. Accordingly, these devices are now being used in numerous laboratories, and studies using oculographic measurements during viewing of complex scenes are on the rise. Nevertheless, there is a considerable number of elementary factors whose effects on eye movement measures are not known. For instance, the influence of image size, viewing time, image content (e.g. natural scenes, pictures of art, regular patterns), stimulus type (e.g. line drawings, photographs, computer-generated images), stimulus repetition, and viewing task (e.g. free viewing vs. recognition) has not been sufficiently investigated (for reviews see Findlay & Gilchrist, 2003; Henderson & Ferreira, 2004; Henderson & Hollingworth, 1998). These, and possibly more, factors could all produce main effects as well as interact with each other in complex ways to influence basic and derived eye movement measures such as fixation duration, saccade amplitude, spatial distribution of fixations, the occurrence of re-fixations, similarity between fixation patterns etc.

This thesis is a continuation of my work which begun with the investigation of the influence of colour on oculomotor behaviour during complex image viewing (von Wartburg et al., 2005). I will focus on four aspects. (1) The influence of image size on saccadic measures will be investigated. (2) New data will be presented concerning the old issue how viewing task influences viewing behaviour. (3) It often seems desirable to repeatedly present images under different conditions. It will be analysed how stimulus repetition might influence – and thus confound – results of image viewing experiments. (4) The temporal course of various measures acquired in the two experiments will also be evaluated and discussed, thereby supplying additional empirical evidence to long-standing questions concerning the change of oculomotor measures over time.

In Chapter 2, an introduction to the basic components of visual exploration behaviour will be given. First, eye movements and visual attention will be discussed separately, followed by an outline of the central aspects and theories concerning their interactions. Finally, a subchapter will be dedicated to questions regarding spatial fixation patterns during scene perception. Chapter 3 reports my experimental work investigating the effects of image size on saccade amplitudes during scene perception. In Chapter 4, a second experiment is described, studying the influence of two important factors oculomotor behaviour, i.e. viewing task and strategies on the one hand, and repeated presentation of the same stimuli on the other. Finally, in Chapter 5, the results will be summarised and discussed in a wider context, in the attempt to integrate the current findings with previous theories and findings, and discuss their relevance in methodological terms.

2 General theoretical part

2.1 Basic components of visual exploration behaviour

In this chapter, the basic components of visual exploration behaviour will be discussed. First, eye movements and visual attention will be considered separately. Second, an integration of the two topics will be attempted. Third, the discussion will be broadened to cover viewing of natural scene images.

2.1.1 Eye movements

The human brain is a powerful cognitive system, and much of its success in enabling its bearer to survive in a complex and potentially dangerous environment lies in the fact that it has sophisticated input systems at its command, the most important of which is the visual system. All human sense systems are tremendously powerful, but the visual system possesses a unique combination of high reach, spatial resolution, directional sensitivity, and a wider range of discernible stimulus dimensions than any of the other senses. This informational richness is too complex for our visuo-cognitive machinery to process all available information in "real time", thus highly specialised structures and mechanisms evolved in order to enable the individual to perform high-resolution analysis of fine detail at a selected location, and keep track of the visual surroundings at the same time.

2.1.1.1 Retinal anatomy

The light-sensitive receptive layer at the back of the human eye is equipped with two kinds of photoreceptors. The *rods* are oblong, thin, cylindrical cells, which are highly sensitive to light and are thus responsible for scotopic vision. Their maximum sensitivity is to light with a wavelength of 500 nm. The *cones* are shorter, thicker, slightly tapered cells, which are less sensitive than the rods and subserve photopic vision. They come in three versions with maximum sensitivities to red, green, and blue light, and provide the input for colour vision (Goldstein, 1999). Overall, the retina can roughly be seen as having a dual nature: A small, central area supporting high resolution colour vision, and the surrounding retinal periphery of almost 180 degrees of low resolution, but highly sensitive achromatic vision which is especially good at detecting motion and flicker. In detail, the retinal structure is made up of several concentric areas. The *fovea* measures approx. 0.6 mm in diameter, or 2° of visual angle, and is defined by the fact that its receptive layer only contains cones, but no rods. The *foveola* is situated in the centre of the fovea and measures approx. 0.3 mm or 1°. Being free of capillaries, it is the area with the highest resolution. The fovea is surrounded by the *parafovea* (approx. 1.5 mm or 5°), where the cones are interspersed with a small number of rods (Liversedge & Findlay, 2000). The rest of the retina is called the *periphery*, where the rods prevail. The density of cones quickly diminishes towards the outside, and so does spatial resolution. The overall distribution of rods and cones across the retina is illustrated in Fig. 1, and Fig. 2 shows the foveal cone density in more detail.

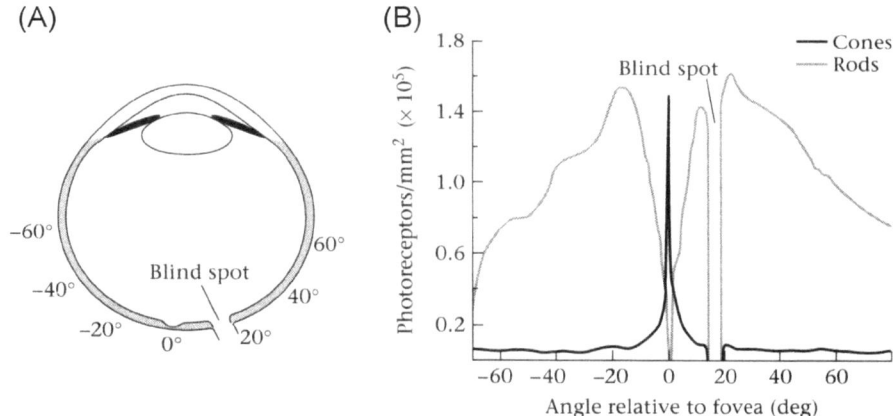

Fig. 1: Retinal structure. (A) Degrees of visual angle relative to the position of the fovea (left eye). (B) Distribution of rods and cones across the retina (from Wandell, 1995).

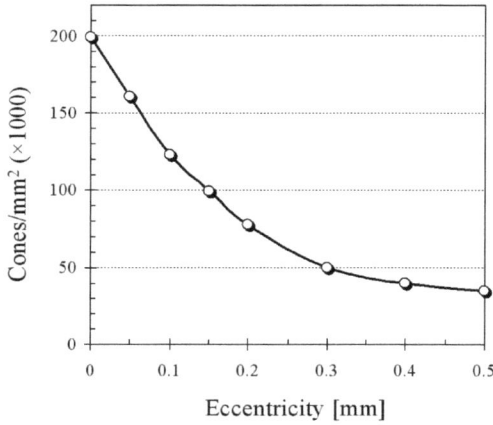

Fig. 2: Foveal cone density, mean values of seven human individuals (drawn after the data provided by Curcio, Sloan, Kalina, & Hendrickson, 1990)

The differential distribution of rods and cones is further accentuated by the "wiring" of retinal ganglion cells: Roughly speaking, very few receptor cells are connected with a single ganglion cell in the fovea, while in the retinal periphery, a ganglion cell bundles the outputs of hundreds of receptors. Accordingly, foveal information is represented by disproportionately larger areas of the lateral geniculate nucleus and cortical visual areas like V1 and higher (e.g. Wandell, 1995).

The names of the different retinal regions and their dimensions, however, often considerably differ between textbooks. For example, Wandell (1995) subsumes all three concentric regions under the term „fovea", whereas for Liversedge and Findlay (2000), only the rod-free part is named „fovea", with a flanking „parafovea". As for the diameter of the latter, values between 5° (Liversedge & Findlay, 2000) and 10° (Rayner, 1984) can be found.

From a functional viewpoint, the definitions of retinal areas are arbitrary, as there is a *continuous* decrease of resolution from the centre to the periphery. Thus, how large an area is sampled for higher visual processing during a single fixation is not strictly determined by the retinal structure. The region from which participants can obtain useful information during a fixation is often referred to as *perceptual span* or *useful field of view*, the estimated size of which largely differs depending on viewing task and research methods. Based on reading studies, the ability to process *fine* detail and thus leading to semantic identification of words is thought to be limited to the central *two* degrees making up the foveal region (e.g. Rayner & Pollatsek, 1992). For pictures of complex naturalistic scenes, the region from which semantic information can be obtained seems to be slightly larger, perhaps three to five degrees (e.g. Nelson & Loftus, 1980). Saida and Ikeda (1979) used a moving window technique, i.e. only a defined area around the participants' current fixation was shown in high resolution while they were performing an image perception task. They found a serious deterioration in the perception of pictures as the window was limited to an area of $3.3 \times 3.3°$ around the fovea, while performance improved with larger fields, and reached asymptote only when it was about 50 % of the entire pattern size.

Considering these facts about retinal anatomy, the essential role of eye movements in human vision becomes clear. If you are at a party and want to find a friend of yours among the guests, it will not suffice to try to find her "out of the corner of the eye"; you will have to let your gaze wander from face to face. That means, in order to take advantage of the high resolution at the centre of the fovea, the eyes must be moved in a way that the to-be-explored parts of the visual field precisely impinge on the fovea; this process is called *foveation*. The information from the retinal periphery does not provide enough detail for fine-grained object recognition, but is, for instance, good enough to keep track of the position or movement of objects previously identified by direct fixation, or to spot new objects that enter the visual field. Moreover, to be efficient, the foveation targets have to be selected as intelligently as possible. Information from the retinal periphery plays an essential role in this selection process.

2.1.1.2 Saccades and fixations

Two main phases of eye movement behaviour are relatively easy to discern. The eyes are rotated towards the foveation targets by rapid eye movements called *saccades*. While a saccade is being executed, virtually no utilisable information is available to the brain, due to a combination of visual masking and central suppression (Matin, 1974)[1]. During foveation of the selected image location, the eyes rest relatively still for a certain time; this phase is called *fixation*. During this time, the primary visual cortex V1 and higher cortical areas are provided with information via the geniculo-striate pathway. At the same time, the superior colliculi of the midbrain and subsequent neural structures receive visual information through a parallel pathway; they play an crucial role in the detection and selection of potential foveation targets. The sequence of saccades and fixations has often been referred to as *scanpath*, even though the original definition by Noton and Stark (1971a, 1971b) rather meant a fixed and often recurring sequence which is characteristic for a certain person and pattern – usually during the initial viewing phase – and not the complete sequence of eye movements. Moreover, the scanpath in its original coining was thought to be linked to memory in image recognition tasks, a link which did not survive close empirical scrutiny. To avoid confusion, I will adopt the term "scan pattern" – as proposed by Henderson (2003) – to refer to an observed sequence of fixations and saccades on a scene.

In physiological research, the focus usually lies on the *saccades*, and the fixations are regarded as mere *saccade latencies*. In cognitive research, the *fixations* are at the centre of interest, because the cognitive processing requirements are assumed to determine the fixated locations as well as fixation duration to a large extent, and are taken as indicators of the deployment of visual attention. In the experiments reported in this thesis, both fixations and saccades are of interest; therefore two separate chapters will be dedicated to their properties hereafter. The anatomy and function of their neural control systems will not be discussed in-depth, because the current research is purely psychophysical and thus not suited to provide data capable of differentiation in neuroanatomical terms. But first, a brief outline of the remaining types of eye movements will be given, for fixations and saccades are by far not the whole story.

2.1.1.3 Other types of eye movements

Bringing targets onto the fovea it is not the only task of the oculomotor system. It is equally important to hold the target *steady* during the phase of information extraction, because both the participant and the foveated object might move. For slow relative movements, this is achieved by *smooth pursuit* movements, which means that the eyes continuously "follow" the object and keep it on the fovea. Actually, there are two mechanisms involved. Movements of the object alone are corrected by a following movement merely based on visual information. If head and body motion is involved, the *vestibulo-optic reflex* (VOR) comes into play, which receives additional information from the vestibular system, allowing for more efficient pursuit. Smooth pursuit movements are much slower than saccades, but

---

[1] as reported by (Henderson & Hollingworth, 1998)

can be initiated faster (less than 50 ms) provided the stimuli have high contrast (Schiller, 1998).

If the differential speed is too high to be corrected by pursuit movements or changes in an unpredictable manner, the system has to recourse to a mechanism called *nystagmus*[2]. The typical nystagmus movement consists of a sequence of smooth pursuit movements alternating with short saccades to "catch up" with the stimulus when it moves too fast to be tracked by the pursuit system. Similar to smooth pursuit, the movements generated when the objects move and the person is stationary are termed *optokinetic nystagmus*; if the person moves, *vestibular nystagmus* takes over (Schiller, 1998).

Besides the *conjugate* saccadic and pursuit movements (i.e. movements in the same direction), there is another type of eye movement called *vergence*, in the course of which the eyes are rotated in *opposite* directions. This is necessary for achieving binocular fusion when we move our eyes from a distant object to a nearer one, or vice versa, along the same line of sight.

Three more types of eye movements not directly coupled to the foveation mechanisms should be mentioned here, because they are important for the definition of fixations during eye movement measurements. On the one hand, a *microtremor* with an amplitude of 20-40 seconds of arc (about 1 to 1.5 times the diameter of foveal cones) and a relatively high frequency of 70-90 Hz occurs during any fixation (Yarbus, 1967). This movement appears stochastic and is not correlated between the eyes. On the other hand, slow back-and-forth deviations from a foveated target, which are called *drift*, as well as corrective microsaccades, can be observed. Contrary to the microtremor, these movements are conjugated. All those movements measure well below one degree in amplitude and can therefore hardly be observed with the naked eye, but measured with precise equipment. The purpose of them is not entirely clear, but if they are suppressed by suitable methods, the so-called "Troxler effect" occurs: A retinal image stabilised in such a way disappears for phenomenal perception after a short time. Therefore it has often been suggested that the transient reaction components of phasic visual neurons fade, and the sustained activity of tonic neurons alone does not suffice to support the perception of contrast discontinuities. So it can be argued that drift and microtremor stand in the service of sustaining stationary vision during foveation (Fischer, 1999).

---

[2] Here, I only refer to *physiological* nystagmus; there are many pathological forms of nystagmus that are not discussed.

2.1.2 Eye movement control models

The question whether eye movements in complex tasks like reading, visual search, or scene viewing are controlled independently from higher cognitive processes has been a matter of much debate and research. The proposed models can be grouped into three conceptual views (Rayner, 1984):

(1) *Global control models* posit that the details of ongoing cognitive processes and eye movements have almost no relationship and that fixation rate largely depends on the participants' competence. The variation of saccadic amplitude and fixation duration is explained as random fluctuations around the pre-set rate. But, as Rayner (1984, p. 78) comments, "… this model is now viewed as something of a straw man as numerous experimental demonstrations have documented a relationship between eye movement characteristics and cognitive processes".

(2) *Indirect control models*: This class of models views eye movements as merely indirectly controlled by cognitive processes, either via memory buffers, or pre-programming mechanisms. According to this view, the processing time needed for encoding the foveal stimuli is too long to have an immediate impact on movement measures.

(3) *Direct control models* suggest that the time the eye remains fixated, as well as the targets of the next saccadic movement, are directly controlled by the information extracted during the current fixation. At present, most evidence, mainly from reading studies, is consistent with this view, but a mixed model of direct and indirect control might be more appropriate to accommodate all evidence.

Given the multitude of different eye movements, another central question arises: Are at least some of these movements controlled independently from others, or should they be seen as tightly interlocked parts of a complex system? As for saccades and fixations, Rayner (1984) concludes that fixation duration and saccade amplitude might be controlled by different mechanisms, given that they can be manipulated independently by means of suitable experimental conditions. The factors influencing these two central measures will be discussed in the remainder of this chapter.

2.1.3 Saccades

As introduced above, the most important type of eye movement in the service of foveation are the saccades. They differ from other types of eye movements by their temporal properties. The defining characteristics are a high initial acceleration of up to $30,000°/s^2$ and a high peak velocity of 400-600 °/s for saccade amplitudes larger than 20°; shorter saccades have correspondingly smaller acceleration and peak velocity values (Cavegn, 1994). These high-speed properties are useful because they minimise the time of disturbed data and potentially maximise the time available for data extraction while the eyes are still. Contrary to smooth pursuit, saccades are *ballistic* in nature, i.e. once initiated, they cannot be interrupted, nor can they change their direction "mid-flight".

The fundamental nature of saccadic movements is nicely illustrated by a case of congenital ophthalmoplegia reported by Gilchrist, Brown, and Findlay (1997). Their patient, a 21 year-old woman, suffers from a complete paralysis of all eye muscles, which means she is completely unable to move their eyes, so she has to move her head in order to bring the objects of interest onto the fovea. This seems to be a dire fate, yet in her daily life she seems not handicapped at all. At a closer look, the movements of her head closely resemble saccadic eye movements. Given the head's mass, it is amazing that the brain does not prefer any other strategy (e.g. smooth sampling), which is a strong indication for the usefulness of saccades. It seems that saccadic movements are the optimal scanning strategy for the brain's processing needs, irrespective of whether they are performed with the eyes or the head.

Saccades are often categorised along their latencies, i.e. the time from the onset of a critical stimulus until the eyes begin to move. Such measurements are performed with very simple stimulus configurations, typically a central dot which a participant has to fixate until it disappears and, after a short delay, another dot appears to the left or right of it, which the participants are required to fixate. This is the so-called *gap paradigm*, as first described by Saslow (1967). The measured latencies typically show a multimodal distribution, and the different peaks are taken as indication of different saccade types. Based on findings by Fischer (1999) and Cavegn (1994), four types can be differentiated:

- Anticipatory saccades have latencies shorter than 80 ms, and are usually not directed towards the target. Saccades with latencies of less than 50 ms are not visually triggered.
- Express saccades occur after latencies of 80-120 ms. They are reflexive in nature.
- Fast regular saccades have latencies of 140-170 ms.
- Slow regular saccades are all remaining saccades with latencies of more than 180 ms.

That such a multimodal pattern can be observed in different trials under identical stimulus conditions is surprising, because more or less identical latencies should occur if only stimulus perception and saccade programming were involved. Obviously, the saccadic reaction time seems to be dependent on certain other internal states of the cognitive system. However, as Fischer (1999) concedes, such clear multimodal distributions cannot be observed in all cases and not in every participant. Moreover, it has to be kept in mind that these data are based on special experimental set-ups which are characterised by near stimulus deprivation properties, as they are usually performed in a completely darkened room, and it is not clear whether the above saccade categories are meaningful in other situations like complex image viewing or natural vision.

If we consider functional aspects, three types of saccades can be distinguished (Pierrot-Deseilligny, Ploner, Müri, Gaymard, & Rivaud-Péchoux, 2002):

Reflexive saccades are *externally* triggered by the sudden appearance or change of a visual stimulus in the periphery, and are therefore also termed *reflexive visually guided saccades*. Reflexive saccades can also be triggered via other modalities, for example by sounds.

Intentional or voluntary saccades are *internally* triggered towards different kinds of targets. If they are directed towards targets that are already present, they are called *intentional visually guided saccades*. *Predictive saccades* are made towards a location where the appearance of a target is expected, and *memory-guided saccades* are performed towards a memorised location where a stimulus has been shown some time before. In certain experimental set-ups, participants are requested to perform so-called *antisaccades*, i.e. saccades towards a location in the same distance but opposite direction of where a stimulus is shown.

Spontaneous saccades are not performed in response to a visual stimulus, but for instance in complete darkness, or during speech.

But again, this saccade typology is based on experiments with similarly unnatural conditions as the gap paradigm outlined above, and might not be indiscriminately applicable to complex stimuli like natural scenes. For example, it is not clear whether saccades made during the initial phase of scene viewing are completely voluntary, if voluntary is understood as being an act of conscious decision making, for their control seems to be largely pre-attentional, although not reflexive in nature. Throughout this thesis, intentional visually guided saccades are of central interest, because with static stimuli as they were used in the experiments, no reflexive saccades are expected. Nevertheless, a small number of express saccades might be observed, because they sometimes occur as *corrective* or *secondary saccades*. Carpenter (1988) noted that saccades of 5-10° are remarkably accurate, while larger saccades often fall short of their intended targets by a fair amount, and shorter saccades frequently exhibit an overshoot. In order to correct for such errors, corrective saccades occur. Depending on the size of the error, these corrective saccades can either be regular (slow or fast ones), if they are small and the final target lies within the "parafoveal deadzone for express saccades", or express if they are larger (Fischer, 1998).

Only sparse information is available on the factors influencing saccade *amplitude*. Over a wide range, it does not seem to be more difficult for the system to make long saccades rather than shorter ones. Therefore it would be uneconomical to move the eye towards a target by several small saccades instead of a single long one, due to the necessary stop/restart latencies between two saccades (see Chapter 2.1.4). Basically, saccade amplitude seems to be determined mainly by stimulus structure and viewing task. Rayner (1984), for instance, reported a mean saccadic amplitude of 2° in reading and 4° in picture perception. Antes (1974) found that saccade amplitudes continually decreased over a viewing period of 20 s. In simple tasks like searching and reading, it is relatively clear where the eyes have to go, which in turn determines saccade amplitude; it seems to be more complicated in scene perception tasks. These issues will be considered in Experiment 1, therefore please refer to Chapter 3 for an in-depth discussion of this topic.

2.1.4 Fixations

The discussion of fixations under the heading of eye *movements* might seem a contradiction, for a fixation could be conceived as the *absence* of movement. But in fact, the term "fixation" is some kind of a misnomer. In the experimental situation with stationary point targets, the eye can be held relatively stable on a certain location ($\pm 0.25°$); only the microtremor described above causes a small amount of "smear". In natural vision, there is no requirement for such fixity of gaze. In addition to the microtremor, micro-saccades, micro-drifts, micro-smooth pursuit movements, micro-VOR movements, and small vergence movements also occur during the time a certain stimulus area is held "steady" on the fovea (Stark & Choi, 1996). In cognitive research, those small movements are usually assumed to be "noise", which is of course meaningful in a functional sense, for these movements stand in the service of stabilising the foveated image parts and sustaining stationary vision.

In the normal course of vision, a myriad of potential targets become available after each shift of the retinal image by a saccade, prompting the saccade system to select one of them and program the next saccade. Therefore, it must not be overlooked that there has to be another major oculomotor function *preventing* unwanted movements. Such considerations gave rise to the concept of a *fixation system*, a separate neural subsystem that serves to prevent the eye from moving for certain periods of time. Fixation is conceived as an active process that inhibits saccade generation while the information on the fovea is being analysed. Based on different experiments – basically variations of the standard "gap paradigm" introduced above – Fischer (1998) concluded that the fixation system can be activated both by visual stimuli close to the fovea as well as by the intention to fixate. There are impressive neuropsychological findings that might corroborate the idea of a separate neural system: In patients with parietal lesions, over-functions as well as under-functions of the fixation system have been found (e.g. Balint, 1909)[3]. While some patients suffer from a *lowered* fixation activity, others show deficits in initiating saccades, which might be due to an *over-function* of the fixation system.

To sum up, the following picture of the functions of the fixation system can be sketched: The fixation system can be activated voluntarily on the one hand, by the intention to keep the eye "on spot", but also by stimulus attributes present in the foveal and parafoveal region

---

[3] as reported in Fischer (1998)

and their processing needs on the other hand. This, in turn, inhibits saccade generation. For a saccade to be generated, the fixation system has to be *disengaged*. This event can be triggered either in a reflex-like manner due to the onset of extrafoveal visual input, or voluntarily.

The most important fixation measure for cognitive research is **fixation duration**. The body of data amassed so far indicates that not only attributes of the oculomotor system *per se* affect fixation duration, but cognitive factors have a strong impact as well. Salthouse and Ellis (1980) postulated a two-component model. The first, **oculomotor component** called *minimum pause time* can be imagined as the time necessary to stop and re-start the eyes, which even occurs when no cognitive operation whatsoever is expected from the participants. It has a mean duration of approx. 200 ms and is only affected by prior movement amplitude, in a magnitude of approx. 6-8 ms per degree of prior movement. There is no significant effect of following saccade amplitude or target complexity. This pause time has to be logically discerned from oculomotor reaction time, for in the case at hand no response was expected, and there was no start signal. Rather, it could be likened to some kind of physiological refractory period necessary for recovering from the preceding eye movement, because programming of the following saccade alone probably takes less time, as Loftus (1981) reported; he found that his participants were able to intelligently program a new saccade even based on an exposure duration as low as 50 ms. Shorter exposures, however, were not sufficient to carry out the peripheral scanning necessary to determine where a subsequent fixation in the picture should be, so they appeared random. However, saccade programming time seems to be largely dependent on experimental settings. For instance, Irwin, Colcombe, Kramer, and Hahn (2000) report typical values of 150 ms for programming of another saccade.

The second, **cognitive component** adds a variable time to the minimum pause time, resulting in the total fixation duration. With a definition of "processing time" as the duration in which 95 % of the stimuli can be processed correctly, Salthouse and Ellis (1980) found that only about 100 ms are needed to process letter stimuli in a vowel vs. consonant decision task. However, various stimulus factors as well as sequential effects among fixations can increase or decrease processing time significantly. It should be noted that the model of Salthouse and Ellis postulates two *independent* determinants, that means it is neutral with respect to whether the minimum pause time and cognitive processing components operate in a serial or parallel fashion. For instance, it may be that the duration of a fixation is not influenced by the time necessary for stimulus processing unless the latter exceeds the minimum pause time. In a scene perception experiment, van Diepen, DeGraef, and d'Ydevalle (1995) found that most of the foveal information can be encoded within an interval of 45-75 ms. They concluded that only the first part of a fixation is necessary to actually encode the object. Nevertheless, the appearance of a mask always tended to increase average fixation duration, even when it was presented late during the fixation. Moreover, it has to be pointed out that all those values are *mean* values. Every participant occasionally produces very short and very long fixations, and some participants even exhibit a *mean* fixation duration of less than 150 ms (Salthouse & Ellis, 1980).

While minimum pause time is relatively constant, the time to encode some feature of the stimulus seems to be highly variable. There are a number of publications which hint at different factors that influence fixation duration in complex visual tasks, but as Henderson and Hollingworth (1998) conclude, the evidence supporting any of these factors is relatively sparse. First, there are several factors affecting **top-down aspects**, i.e. the plan or strategy one adopts to solve a certain task. The most important variable is probably *viewing task*, with its direct impact on strategies. Again, a further differentiation is necessary. (1) Tasks can be completely different in a sense that *stimuli* as well as *instructions* differ. In a review article, Rayner and Pollatsek (1992) conclude that in image perception, the average fixation duration is longer than in reading, perhaps because more useful information can be taken in a single glance during image perception. An alternative explanation might be that the cognitive system needs more time to extract the desired information from a typically cluttered surrounding of complex scenes, while there is practically no "noise" in a normal text display. During visual search, fixation duration lies somewhere in-between. As in reading, the objects are usually more highly constrained in the typical search task display than in natural scenes. (2) Tasks can also be different in a sense that the same stimuli are shown under different task instructions. Rayner and Pollatsek (1992) suggested that if a scene is shown to participants who are required to *search* for a target object, they will most likely show longer fixations than other participants who are asked to look at the scene, but in anticipation of a *recognition memory test*. In a similar fashion, participants will exhibit slightly longer fixations if they are asked to *memorise* all the objects they can see, compared to *simply looking* at the scene in order to extract the gist of it. This is attributed to the fact that in the former case additional time is needed for memorisation. However, this seems to depend on the exact stimulus and task properties. The influence of viewing task on various oculomotor measures will be investigated in Experiment 2, so please refer to Chapter 4 for an in-depth discussion of these issues.

As for the **bottom-up aspects**, the current data situation is not much better. To start with, I recently investigated the influence of colour on oculomotor behaviour during image viewing (von Wartburg et al., 2005). The aim of this study was to investigate how oculomotor behaviour depends on the availability of colour information in pictorial stimuli. Forty participants viewed complex images in colour or grey-scale, while their eye movements were recorded. The main finding contradicted the widely accepted theory that fixations become longer with increasing complexity. Although the inclusion of colour information actually increases the complexity of an image, fixations on colour images were shorter than on their grey-scale versions. This suggests that colour enhances *discriminability* and thus affects low-level perceptual processing.

Henderson and Hollingworth (1998) presented evidence for effects of different *image types*: Comparing photographs, line drawings of the same scenes, and 3D rendered images of similar scenes, they found that participants made slightly but reliably longer fixations on photos than on both other – more artificial – versions. As the above reported results concerning colour, this result can also be framed in terms of *discriminability*, as the information in a line drawing is represented more abstractly and free of "visual noise", and therefore more easily discriminable than a typical photograph. The same is valid for the computer-generated images; typically, such images are also less afflicted with visual noise. It could be speculated that just adding noise to an image would prolong fixations in the same manner. In a similar fashion, fixation duration was found to increase with decreasing stimulus luminance (Loftus, 1985), and if a picture is more densely packed with information, fixation duration will increase as well (Rayner, 1984; Rayner & Pollatsek, 1992).

Right from the beginning of research in this area, it has been speculated that viewing behaviour changes over time. Buswell (1935) reported two kinds of eye movement patterns: An initial "survey scan" phase in which the eyes move over the image with relatively short fixations and long saccades, and a "detailed looking" phase with longer fixations which were more concentrated on specific areas, usually appearing after the survey scan. In a study comparing normal participants and patients with various brain lesions, a similar "preliminary orienting period" was found (Karpov, Luria, & Yarbus, 1968). In particular, the pattern was disturbed in patients with bilateral frontal lesions. Other data reported by

Antes (1974) basically confirmed these findings: Over a viewing period of 20 s, mean fixation duration was found to increase continually from 215 to 310 ms, while saccade amplitude decreased from 4.1 to 3.4°. Therefore, the course of events is not seen as two sharply demarcated phases but rather as a continuum between two extremes, from surveying informative aspects of the entire picture to the examination of less informative details. This issue will be considered in both experiments reported in this thesis.

In conclusion, many questions concerning the determinants of fixation duration remain open. This thesis is designed to give a more complete picture of the most important issues in image perception and eye movement studies, in theoretical as well as methodological terms.

2.2 Visual attention

A multitude of definitions, metaphors, and aspects of attention have been forwarded during more than a century. Attention has been conceptualised as filter, skill, selective attenuator, spotlight, or zoom lens, to name only a few (Wright & Ward, 1998). Most of them are also applicable to *visual* attention. Basically, this thesis was not intended to provide evidence helping to resolve which flavour of attention is the most appropriate. Nevertheless, questions concerning visual attention might be relevant for the interpretation of the results, as attentional processes are critically involved in the control of eye movements and the selection of foveation targets.

2.2.1 General considerations

Unfortunately, there is no unequivocal and generally accepted definition of the term "visual attention". Nevertheless, we all know what to do when we are instructed to "pay attention" to a certain location in the visual field, and our efforts to do so usually lead to corresponding changes in our performance. Often, a definition starts with the introspective observation that we are obviously able to turn our attention to an object, even without directly looking at it. Concepts of visual attention – or, more precisely, *selective* visual attention – are usually substantiated by experimental results that demonstrate the possibility to somehow prefer certain parts of the visual environment over another, leading to *prioritised processing* of the pieces of visual information attended to. This can, for instance, be demonstrated by comparing the performance in two identical perceptive tasks, but with differential allocation of attention. A typical experiment is a lateralised discrimination task with leading cues to guide participants' attention towards one side of the display. If the cue is correct, i.e. if it correctly points to the side where the to-be-discriminated stimulus appears a few moments later, reaction times will be shorter than with misleading cues.

But why do we have the ability to selectively attend to different aspects of visual information? In the context of computational modelling, the following functional definition has been widely accepted (Itti & Koch, 2001): The most important function of selective visual attention is to rapidly direct our gaze towards "objects of interest" in our visual environment. The ability to rapidly orientate towards salient objects in a cluttered visual scene has important evolutionary significance, for it enables the organism to spot food, predators, and potential mates efficiently. Another aspect, which is particularly relevant with regard to computational modelling, is that attention protects the visual system from overload. In this sense, attention is a selection process which implements an information-processing bottleneck allowing only a small part of the incoming sensory information to reach short-term memory and visual awareness. Because of the capacity limitations of the cognitive system, it is impossible to fully process the massive sensory input, thus a serial, capacity-limited strategy has evolved that achieves virtual real-time performance in most everyday situations. The visual scenery is thereby dissected into small pieces that can be analysed more easily by higher visual functions such as object recognition. This problem has clear parallels in computer vision. It is easy to imagine the implementation of a form analysis and recognition algorithm for a specific location in an image, but it is nearly impossible to apply such routines to the whole visual field at once, because this would quickly lead to a combinatory explosion and therfore exhaust the computational resources. As in human perception, a "piecemeal approach" is often attempted, by breaking down the analysis tasks into subtasks that are computationally less demanding.

The "classical view" of visual attention is to imagine it as a spotlight that moves over the visual scene, highlighting certain parts here and there. The two central aspects of the spotlight metaphor are the size of the "beam" and the way how it moves. The first question is how large the "illuminated" area is, and whether it is constant over time or whether it can be widened or focused. The second aspect concerns the manner of the displacement when it changes its location. The "analogue spotlight" model assumes that it moves in a continuous fashion, or wanders from location A to location B, thereby also highlighting the locations along a straight path from A to B for a short time. Contrary to this, the "discrete spotlight" model imagines the spotlight to jump from A to B without delay. The "analogue zoom lens" model assumes that when processing of location A is finished, the spotlight defocuses and "illuminates" almost the whole scene, and then shrinks and focuses on location B (for an overview see Wright & Ward, 1998). Recently, Brefczynski and DeYoe (1999) presented evidence that visual attention might literally "highlight" the regions attended to. In a functional magnetic resonance imaging (fMRI) study, they had their participants either view a stimulus at a certain location, or required them to direct the attention to the same location, but without the stimulus being shown. The activity pattern in the visual cortex evoked by the purely attentive operation was identical with the activity due to viewing the targets at the same position, especially in V4.

As the present experiments are aimed at the question *where* and *how long* visual attention is directed in the course of image viewing, and not what it exactly is and how it works, we can content ourselves with a minimal definition: Basically, visual attention refers to a class of mechanisms that select incoming information for subsequent processing from a particular set of locations in space and a certain interval in time. Throughout this thesis, visual attention is conceptualised as the ability of the visual system to detect and select potentially relevant parts of a scene in order to facilitate processing on higher levels. Or, as Henderson (1992b, p. 260) put it, visual attention is "the selective use of information from one region of the visual field at the expense of other regions of the visual field." In the context of my thesis, the *shifts* of visual attention are of central interest, or in other words the changes in the spatial location to which we attend. They usually accompany eye movements but can also occur independently of eye fixation. Thus the deployment of visual attention will first be discussed independently of eye movements, and the relation to eye movements will be considered afterwards.

2.2.2 A two-component model of visual attention

In recent time, evidence from psychology and computational neuroscience has accumulated that favours a two-component model. The mechanisms controlling visual attention can be framed in terms of the top-down vs. bottom-up dichotomy. This model should not be seen as a two-*stage* model with components operating serially, but as two major components which operate in parallel and, perhaps, to a certain extent independently. In the following, two main concepts will be used. **Top-down attentional control** embraces mechanisms that rely on concept driven processes like volitional decisions, schemas, familiarity, task-dependent strategies, and the like. **Bottom-up attentional control** encompasses stimulus- or data-driven processes that operate on stimulus features like contrast, colour, saliency, discriminability, etc. They will be considered in turn.

The distinction between top-down and bottom-up processing stems from computer science (Stevens & Rumelhart, 1975); its application to visual attention traces back to Treisman and Gelade (1980). Their *feature integration theory* postulates that a visual array is represented in separate "feature maps" for different aspects of visual information, e.g. brightness, colour, orientation of line segments, or other form parameters like "curvature". These feature maps are assumed to be built by parallel bottom-up processes which operate pre-attentively. If an array of elements is presented in a visual search task, and only one feature differs for the target item, attention will immediately be drawn to that item, it will "pop out" (see Fig. 3a). However large the array is, no previous scanning movements are necessary, as long as the critical feature can be discriminated within the presented stimulus array at all, given the acuity limits and lateral interference at the affected retinal areas. In a similar fashion, attention can be drawn reflexively by peripheral stimuli, especially if they move or show a fast onset or change. This kind of selection is based on iconic, or appearance-based, scene representations, that is no categorial, or schema-based, decisions are involved. Therefore, it is also called data driven or exogenous selection. But if the target item is defined by two or more features (*conjunctive targets*), participants must revert to a serial search strategy (see Fig. 3b), which is a top-down mechanism. The information has to pass the "bottleneck" of serial processing, therefore search time will increase with the number of items in a linear fashion. In the view of Treisman and Gelade (1980), focused attention can only be positioned at one location at a time, and attention is needed to "bind together" the different stimulus attributes represented separately in the different feature maps.

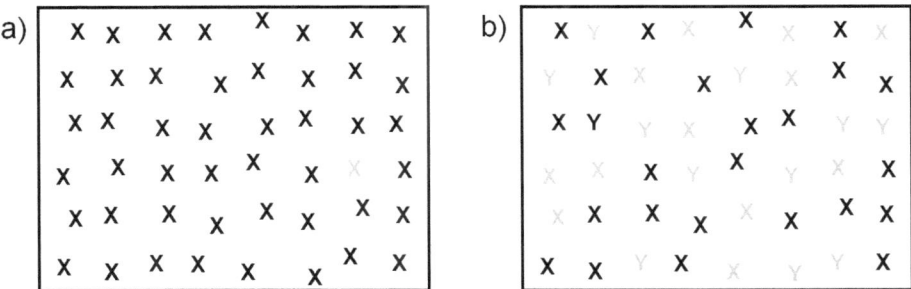

Fig. 3: Examples of search tasks. a) "Pop-out" task: Find the odd one out. b) Conjunctive search: Find the black Y.

There are several aspects of endogenous (concept driven, strategic) control of attention. First, it seems introspectively plausible that visual attention can be guided by central volitional decisions such as goal-dependent search plans or other strategies, or simply the decision to keep attention at a certain location. This means that this component is task-dependent, be it with regard to "overt task" like finding a certain person in a crowd, reading, car driving, or any experimental task that has been explicitly assigned, or be it with more "implicit" tasks which are activated by certain kinds of visual situation, i.e. in viewing landscapes, attention might be deployed differently compared to admiring fine arts in a museum. As all processes involving decisions and strategies, this attentional deployment mechanism is probably controlled by higher, frontal areas of the brain (Itti & Koch, 2001).

A second aspect is how *schemas* can aid efficient visual scanning. Although experienced readers do not have to explicitly control their scanning behaviour in reading, a cognitive schema of how text is arranged on a newspaper page aids efficient reading. In daily situations, "scene schemas" as suggested by different researchers (Antes, Penland, & Metzger, 1981; Biederman, 1972; Henderson, 1992a) might be more important. The *scene schema hypothesis* proposes that "fast scene recognition depends on the early activation of a few scene representations in memory to drive a top-down extraction of information in the noisy input" (Schyns & Oliva, 1994, p. 195). Moreover, scene context has been shown to facilitate object searching if the objects are shown in the appropriate context (e.g. (Biederman, Rabinowitz, Glass, & Stacy, 1974). This illustrates that top-down does not always mean attentive; while the strategies aspect has a large attentive part, schemas usually operate without being recognised. Clearly, those two components should be seen as independent. In contrast to pop-out tasks with only one different item, imagine the typical conjunctive target search task (Fig. 3b). Nobody will look at *every* item serially, but the bottom-up saliency will aid the serial search in a way that only the black items will be closely examined in order to find the black Y.

Inspired by the feature integration theory, a relatively well-specified model of the bottom-up attentional control mechanisms has been developed by research groups from psychology and computer science. Based on publications by Niebur, Itti, Koch, and Ullmann (e.g. Itti & Koch, 2001; Koch & Ullman, 1985; Niebur, Itti, & Koch, 2001), the following picture of the involved processes emerges. Visual input is represented in the retina and in various subcortical and cortical structures. The first step consists of the calculation of simple visual features, which are processed at different stages: in the retina, the superior colliculus, the lateral geniculate nucleus, and early visual cortex (V1-V5, depending on the feature). There are separate maps for certain *elementary features* as suggested by Treisman (1983), like intensity contrast, colour opponency, orientation, direction, speed of movement, and perhaps stereo disparity. For certain features, these maps can be further subdivided into sub-maps for different stimulus dimensions, e.g. the colour map consists of two colour-opponency sub-maps which code for yellow/blue and red/green separately. They might be the neural correlates of the feature maps suggested by Treisman and Gelade (1980). Corroborating their theory and experimental evidence, neuroscientific research showed that neurons on these levels operate pre-attentively and in parallel over the whole visual field. "Pre-attentively" means that those neurons fire vigorously in response to their target stimuli irrespective of whether attention has been directed to that region, and even in an anaesthetised animal.

The crucial step in the construction of these feature maps is a centre-surround evaluation of every feature dimension, which is performed already within the early representations. Not the absolute strength of a feature is important, but contrast with the surroundings. This mechanism is comparable to retinal centre-surround mechanisms. However, it does not only operate on the scale of classical receptive fields, but on various different (and mostly larger) scales. In accordance with Treisman and Gelade (1980), there are no signs for any strong interaction between different visual modalities (i.e. feature dimensions). Otherwise top-down modulation – e.g. learning – might be expected, but obviously, we are not able to learn detection of conjunctive targets to a large degree. So each of these feature maps codes for the conspicuity *within* a single featural dimension. Afterwards, they are combined into a "saliency map", a concept popularised by Koch and Ullman (1985). The saliency map is a topographical map containing the merged local maxima arising from the single feature maps. The different features contribute to perceptual saliency with different strength. This relative weighting is assumed to be modulated by top-down influences (e.g. task strategies) and training. By definition, the momentary maximum on the saliency map is the most "conspicuous" location. In purely computational models, the local maxima are now scanned in decreasing order by means of a competitive "winner-take-all" process in order to select focusing targets. This process incorporates a lateral inhibition mechanism in order to avoid generating several target locations which are too near to each other and therefore would be non-functional, and an inhibition-of-return mechanism preventing early returns to previously attended locations. For a recent description of this model see Ouerhani, von Wartburg, Hügli, and Müri (2004) or Parkhurst, Law, and Niebur (2002). This computational model easily explains a central property of the feature integration theory. If there is a significant difference in only one stimulus dimension (i.e. feature map), only a single peak results in the final saliency map, and so it will be focused immediately (pop-out). If there are more peaks, their locations must be attended to serially.

If the observer directs attention to a certain spatial location, the contents of all associated feature maps are merged into a central representation which itself is non-topographic and more abstract. This corresponds to the "binding" or "gluing" postulated by Treisman and Gelade (1980). So the winner-take-all process only *selects* the location for central processing, but it is not in itself the processing.

In sum, the two-component model can be outlined as follows. There is a pre-attentive, bottom-up component, which quickly processes different visual features in parallel over the whole visual field, and can guide the attentional focus to most salient regions. In addition, a top-down component modulates bottom-up saliency based either on conscious, volitional decisions or by using previous visual experience (e.g. scene schemas) which is activated without conscious effort. This component can be thought of as an increase of stimulus intensity in certain regions.

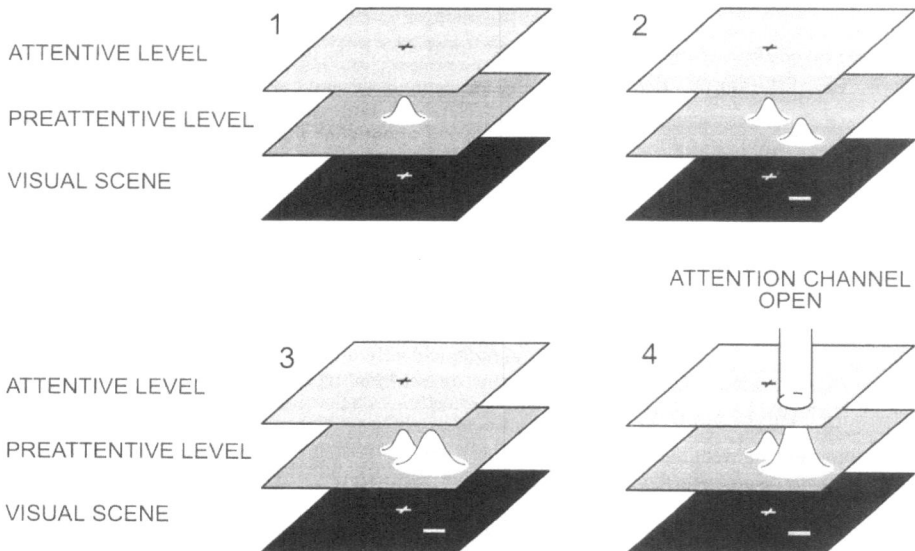

Fig. 4: A simplified version of the activity-distribution model of attention. The top surface symbolises the higher-level representation within which a channel of focused attention can be opened if activity exceeds a threshold level. The peaked distributions are a form of pre-attentive location encoding (adapted from Wright, 1998).

A proposal by LaBerge and Brown (1989) nicely integrates these different aspects in a model called "activity-distribution model of attention". In addition to a pre-attentive, stimulus driven activity map similar to the saliency map concept, they suggest a second map which represents the goal-driven attentional activity. If the summed activity accumulated at a certain location reaches a threshold value, an "attention channel" opens and the stimuli can be processed by higher faculties (Fig. 4). Similar to the discrete spotlight idea, a channel at a new location is opened simultaneously with closing of the channel at the old location.

This theory recently received support from several experiments, which provide evidence that is inconsistent with the different spotlight models, but consistent with the activity-distribution model (LaBerge, Carlson, Williams, & Bunney, 1997; Pratt & Quilty, 2002).

2.3 Visual attention and eye movements

Basically, the discussion how eye movements and visual attention are related is not concluded (for a discussion of this topic see Sanders and Donk (1996). In this chapter, the question whether eye movements can be used to indicate the deployment of visual attention in image viewing will be discussed.

Two terms are central to this topic, covert and overt attention. **Covert attention** designates an attentional shift without a concomitant saccade. Although a tight coupling of eye movements and visual attention is assumed in most situations, there is evidence that it is possible to shift attention to a more peripheral location without letting the eyes follow (e.g. Posner, 1980). This is what we all know when we watch something "out of the corner of the eye". While this mechanism is not very useful in the inspection of a new scene, it can be quite efficient in order to keep track of something interesting that has been identified before. Nevertheless, in normal daily viewing this might be a moot point, since it is so easy to fixate an object of interest. **Overt attention**, on the other hand, is used when a shift of visual attention to a certain target spot is accompanied by an eye movement to the same location. This seems to be the normal case in daily looking behaviour. As Wright and Ward (1998) pointed out, foveation is an integral part of the orienting reflex, and in general any foveating saccades would be non-functional if the attentional focus were independent of them. Given the retinal structure, it would be utterly useless to shift attention to an important spot without making a saccade, or vice versa, for the higher-level processing would be severely constricted due to the reduced resolution beneath the fovea.

Actually, for a normal overt attention shift, two things must be shifted, attention and the eyes, so the question arises whether the target locations for the two shifts are determined independently. Meanwhile, there is overwhelming evidence that this is not the case. Hoffman and Subramaniam (1995) demonstrated that participants cannot attend to one location and move their eyes to a different one. Deubel and Schneider (Deubel & Schneider, 1996; Schneider & Deubel, 1995) strongly argue for an obligatory and selective coupling of saccade programming and visual attention to one common target object. They had their participants make saccades to locations within horizontal letter strings left and right of a central fixation cross, and tachistoscopically presented normal and mirrored "E" letters before the saccade within the surrounding distractors, which the participants had to discriminate. The performance of the participants was best when the discrimination stimulus and saccade target referred to the same object. Interestingly, even when the saccade was inaccurate and a corrective saccade was necessary, the performance was better for the *intended* target location. In conclusion, it was impossible for the participants to direct attention to another location than the one set as saccade target. This clearly argues against the so-called *decoupling hypothesis*, i.e. the ability to direct attention to one target and to program a saccade to another one at the same time. In concordance with these suggestions, Hodgson and Müller (1995) found that, for both reflexive and voluntary saccades, whenever a saccadic eye movement is about to be executed, the direction of visuospatial attention is constrained to be compatible with the direction of the eye movement. Wright and Ward (1998) express this idea in a hybrid model, called the *destination indexing model*. According to this model, some attentional processes are independent of those involved in programming oculomotor movements, but both share a common mechanism that encodes the location of target destinations. *Indexing* is the process by which location information is made available to other visual operations.

In general, there seems to be a broad consensus that the target location of visual attention and eye movements are coupled. The question how an attentional and eye movement shift takes place in detail, however, still is a matter of much debate and research. An early, influential hypothesis is the "premotor theory of attention" (Rizzolatti, Riggio, Dascola, & Umiltá, 1987), which postulated that attentional shifts to a stimulus were made on the basis of eye movement programs to that location. Thus, an eye movement program necessarily preceded any shift of attention. However, evidence provided by Hoffman and Subramaniam (1995) challenges this concept. They found that planning a saccade towards a location improves the detectability of information presented in that location before the saccade. This suggests that performing a saccade involves orienting of attention to that location *prior* to the actual execution of the saccade, and the attentional shift precedes the eye movement. Henderson, Pollatsek, and Rayner (1989) support this notion, and propose a "sequential attention model" with five basic assumptions.

> First, at the beginning of each new eye fixation visual attention is allocated to the stimulus at the center of fixation. Second, attention is re-allocated to a new stimulus when processing of the foveal stimulus is completed. [...] The stimulus to be attended next is determined by relative weights, based on salience of preattentively determined stimulus locations [...]. Third, re-allocation of attention is coincident with two aspects of eye-movement programming: (a) The system begins to program the motor movements necessary to bring the eyes to a new location, and (b) the new locus of attention is

taken to be the location toward which the eyes should be moved. Fourth, the reallocation of attention to a new location gates higher-level analysis at that new location. Fifth, the eyes follow the shift of attention to the attended location following a constant (plus noise) eye movement-programming latency. (Henderson, 1993, p. 42)

Now, it might be objected that what has been discussed seems to speak against the obvious possibility of covert attention shifts. The consensus seems to be that although programming of the attentional and saccadic shift is indeed coupled, the *execution* of the saccade is not compulsory, and the programmed saccade can be withheld voluntarily, which is for instance seen as a missing release of a fixation signal (Deubel & Schneider, 1996).

Based on recent research using search tasks, it can be concluded that, even though covert attentional shifts are possible, most search tasks will be served better with overt eye scanning (Findlay & Gilchrist, 1998; Maioli, Benaglio, Siri, Sosta, & Cappa, 2001); it seems plausible to postulate this for scene image viewing as well. When the eyes are free to move, there is no reason to postulate shifts of visuospatial attention other than those associated with the execution of saccades. In conclusion, it is generally accepted in empirical research that, at least insofar as everyday viewing behaviour is concerned, eye movements and visual attention are tightly coupled. Thus, eye movements – or, more precisely, the location of fixations in particular – can be taken as a reliable indicator of the deployment of visual attention.

2.4 Spatial fixation patterns during scene perception

After having introduced the basic aspects of eye movements and visual attention, the discussion shall be broadened by turning to the specific aspects of oculo-motor behaviour during the exploration of natural scenes. In the previous chapters, basic aspects of fixations and saccades have been introduced; they can be expected to hold for scene perception as well. In the current chapter, only the aspects concerned with the spatial location of fixations, or *topology of fixation*, relative to the stimulus image will be discussed.

2.4.1 Aspects of exploratory behaviour

Various approaches to evaluate and interpret fixation location data have been proposed hitherto. Two main distinctions are whether individual or grouped data are considered and whether the distribution of fixated locations is set in relation to time or not.

Most of the early studies concentrated on **absolute fixation distribution** as acquired from a group of participants, considering the whole viewing period. With this approach, the fixated locations are evaluated as absolute co-ordinates and may be visualised by means of fixation density plots and the like. In order to relate fixation data to specific hypotheses, they are often counted for several "regions of interest" of the image. An interesting method of comparing group data was proposed by Mannan, Ruddock, and Wooding (1995). Their index of similarity allows for comparisons of fixation patterns between or within participants, as well as relating fixation data to image-related locations. Some research groups also differentiated between early and later phases of exploration, thus investigating **fixation distribution over time**, but still based on group data.

Others tried to analyse the individual temporal sequence an observer exhibits during the course of image inspection. The sequence of saccades and fixations is often termed **scanpath**, even though it was initially defined in a more specific sense (Noton & Stark, 1971a), which will be considered shortly. Some years before, Yarbus (1967) noted that

> If the eye movements are recorded for several minutes during perception of an object, the record obtained will clearly show that, when changing its points of fixation, the observer's eye repeatedly returns to the same elements of the picture. Additional time spent on perception is not used to examine the secondary elements, but to re-examine the most important elements. The impression is created that the perception of a picture is usually composed of a series of "cycles," each of which has much in common. (Yarbus, 1967, p. 193)

In a study aimed mainly at the investigation of memory processes in interaction with oculomotor processes, Noton and Stark (1971a, 1971b) found similar "cycles" during visual inspection. They presented five different line drawings of 20° visual angle to six participants, for 20 to 75 s each. In order to force participants to look at the image parts foveally, the visibility of the drawings was reduced by using very thin lines and low illumination, thereby imposing a serial processing strategy upon the participants. For the topic at hand, the important result was that they found a characteristic sequence of fixations for participants and patterns, followed intermittently but repeatedly. These sequences often recurred during recognition trials under the same stimulus conditions applied later, supposedly facilitating stimulus recognition. These properties make up the initial definition of a scanpath: "Such a fixed path, characteristic of a given subject viewing a given pattern, we have termed his 'scanpath' for that pattern" (Noton & Stark, 1971a, p. 310). In this sense, a scanpath is often observed as an initial scanning sequence rather than the complete set of fixated locations, and often recurring over time. But it has to be noted that in Noton and Stark's study, not all stimuli evoked a discernible scanpath, as the authors admitted. Groner, Walder, and Groner (1984) extended the concept, by postulating local and global scanpaths. *Local scanpaths* are consistent patterns of consecutive fixations, thought to be controlled in a bottom-up manner from fixation to fixation. *Global scanpaths* represent the distribution of fixations over the viewing period on a larger timescale; in this sense, they reflect top-down processes such as the search plan or the hypotheses of the participant concerning a certain image. In a face recognition experiment, they found limited evidence for both types of scanpath. Four out of six participants exhibited a significant number of scanpath triplets, and two out of six showed a strong tendency towards global scanpaths. Mannan, Ruddock, and Wooding (1997b, p. 164) could only provide "very limited evidence of fixed sequences of eye movements performed to the same image" within participants, even for two-fixation sequences, which would be equivalent to local scanpaths. However, contrary to the "classic" design, the second presentation was not part of a recognition task. On the other hand, similarity of fixated spatial locations was high when sequence information was disregarded. After all, Pieters, Rosbergen, and Wedel (1999) claimed strong support for predictions derived from scanpath theory as applied to repeated print advertisement viewing, where scanpaths are expected to facilitate the identification of advertisements' contents in repeated exposures. In conclusion, the evidence for scanpaths is limited, and whether scanpaths occur during natural scene perception has not been sufficiently investigated. This issue will be tackled in Experiment 2 (see Chapter 4). To avoid confusion, I will use the general term "scan pattern" for any spatio-temporal sequence of fixations and saccades occurring during viewing of an image, as suggested by Henderson (2003). With the expression "scanpath" I will exclusively refer to spatio-temporal sequences of fixations that *repeatedly* occur during several presentations of the same stimulus image.

2.4.2 What guides visual exploration in scene perception?

To be effective, the attentional system must somehow manage to quickly direct the eyes, which are in most situations coupled to visual attention, to the locations where information critical for the organism's current needs might be found, and reject irrelevant information for reasons of cognitive economy. A random selection mechanism would be very inefficient, as well as a stereotypical scanning strategy that samples any scene in the very same way. So what we have here is the "paradox of intelligent selection": How should a selection mechanism be conceived that is able to select the most important locations without first having to process all available information?

A fundamental question is whether the image areas not fixated – i.e. the light pattern falling outside the fovea – can be processed to the extent that the extracted information can be used to guide the following fixations to "useful" areas. Obviously, this is the case, as human exploratory behaviour is neither random nor systematic in the sense that every scene would be scanned in the same sequence. That is, during the current fixation, not only foveal information is acquired, but peripheral information is used to determine the next location to be attended to. But the relative contribution of the different factors suggested is not yet clear.

2.4.2.1 Bottom-up saliency

According to the saliency map approach (Itti & Koch, 2000, 2001; Koch & Ullman, 1985) outlined in chapter 2.2.2, the deployment of visual attention is partly controlled by bottom-up, image based information. Recently, this theory has been confirmed by experimental studies (Jost, Ouerhani, von Wartburg, Müri, & Hügli, 2005; Ouerhani et al., 2004; Parkhurst et al., 2002). Therefore, it seems plausible to assume that these mechanisms are highly relevant for scene perception in general.

2.4.2.2 Image content and informativeness

When we look at a complex display, do we rather concentrate on meaningful parts of the scene, or do we favour those with high visual saliency? Early research indeed suggested that informative regions obtain more fixations than others. As mentioned before, Buswell (1935) claimed that eye movements of different participants were highly regular and clearly related to the information in the images. For instance, the participants tended to look at people in the image foreground much more often than on the background. The impression that informative and therefore (by definition) important areas receive the majority of fixations was corroborated by other findings. Mackworth and Morandi (1967) presented pictorial stimuli to one group of participants and recorded their fixations on the images. Another group of participants were asked to rate 64 square fragments of those images for informativeness, which was defined in terms of recognisability. The main result was that the "highly informative" regions attracted more fixations, not only in the beginning of viewing, but over the complete ten seconds of presentation, whereas redundant contours, although highly recognisable, received few fixations; areas of mere texture scored even lower. There seemed to be a rapid rejection of unwanted, redundant areas at the beginning of the period of inspection. Alas, only two different images were used, and quite uncommon ones: a picture showing "a pair of eyes behind a crimson mask" and "an astronaut's view of Baja California", so they were relatively impoverished from the point of view of semantic complexity (Friedman & Liebelt, 1976).

A similar method was adopted by Antes (1974). He presented achromatic shaded drawings (mostly from the Thematic Apperception Test (TAT) for ten seconds each. The informativeness ratings were not made for equal-size square pieces, but for "regions of interest" which were determined post hoc based on the fixation densities obtained in the experiment; the criterion was "informational contribution of that unit to the total information conveyed by the picture" (p. 65). Regarding fixation distribution, basically the same results were found: The participants immediately fixated the most informative element. Loftus and Mackworth (1978) used line drawings of a variety of scenes, and swapped single objects between them, so they became inconsistent and therefore "semantically informative" (e.g. a tractor in an underwater scene vs. a octopus in a farm scene). By concentrating on the first fixation on a critical object, they found that participants tended to fixate the inconsistent objects earlier during the course of scene viewing and were more likely to fixate it immediately following the first saccade compared to the same object in the consistent scene. Because the distance of the target objects averaged 6.5 - 8°, these data again suggest that the location of the inconsistent objects could be determined in a single fixation, based on peripheral information.

However, this is surprising, as the area from which semantic information can be gleaned is generally estimated to be much smaller. Indeed, there is a basic problem with most of these studies. There was no adequate control whether the semantic content was correlated with basic visual information; therefore it is difficult to divide the processes into perceptual and cognitive components. And so it comes not entirely unexpected that evidence to the contrary has been found as well. Henderson and colleagues (Henderson & Hollingworth, 1998; Henderson, Weeks, & Hollingworth, 1999) used line drawings produced after photographed indoor scenes, and replaced some of the objects with inconsistent ones. Their data indicate that the eyes are *not* initially driven by peripheral semantic analysis of individual objects, as the inconsistent objects were no more likely to be fixated immediately after the first saccade than any other object. But when the whole exploration phase was evaluated, a clear tendency to fixate the inconsistent/informative objects more often became apparent. The number of fixations participants made in a certain region *after the first fixation* seemed to be dependent on the semantics, and participants tended to return to such regions later on. A recent study found confirming evidence that this applies to full-colour scenes as well: Object deletions or substitutions could not be recognised based on information acquired from the visual periphery (Henderson, Williams, Castelhano, & Falk, 2002). So the preliminary conclusion is that initial fixation placement in a complex, natural scene is not controlled by a peripheral semantic analysis of individual objects in the scene. However, earlier findings may partly be explained by the fact that once an object has been fixated, the eyes tend to return to semantically informative objects more often and fixate them for longer intervals.

2.4.2.3 Scene schemas

After I have discussed several bottom-up influences of visual exploratory behaviour during scene viewing, I now turn to top-down aspects. Regarding natural scenes, it seems probable that the gist of the entire scene can already be extracted during the first fixation (Biederman, 1972, 1981; Biederman et al., 1974). Biederman (1972) as well as Schyns and Oliva (1994) showed that human observers are able to recognise complex natural scenes as fast as isolated objects, in a single glance of 200 to 300 ms, or even less. This is consistent with the aforementioned "scene schema hypothesis", which holds that scenes are represented as clusters of oriented blobs of specific sizes and aspect ratios, organised in particular graphs of spatial relationship. Therefore *scene categories* tend to have distinct and typical spatial organisations of their major components, which could provide the information for a mechanism by which scene schemas are activated (Schyns & Oliva, 1994). The activated schemas might facilitate finding and recognising of objects, by providing guidelines where meaningful objects can be expected with higher probabilities. Empirical evidence for this was provided by Henderson and Hollingworth (1998) whose participants found semantically consistent items in a search task earlier than inconsistent ones, probably because their possible positions were constrained by the scenes. The results suggest a time- and spatial-scale-dependent scene recognition process in which the very first stages rely on scene-specific information and the later stages are object-based. By these means, the scene is divided into informative features and redundant regions "at first sight". A short scan of the complete image is not necessary to this end. Based on these ideas, it is assumed that – in addition to bottom-up saliency – scene schemas provide important guidelines for the selection of saccade targets.

2.4.2.4 Viewing task

The second – and maybe most important – top-down influence is based on the strategies an observer selects, given the concrete task assigned for scene image viewing. As Rayner and Pollatsek (1992) noted, terms like "image viewing" are somewhat fuzzy, and there is no well-defined task of "scene perception". If not given explicit instructions, people search pictures for *meaning* and not for specific targets (Gould, 1976), i.e. they are actively searching for information relevant to current motivations and goals. As demonstrated by Buswell (1935) and Yarbus (1967), the assigned task not only influences basic oculomotor measures, but also the *location* of fixations, i.e. the scanning strategy in spatial terms. As this topic is of central interest, it will be further discussed and empirically investigated in Experiment 2 (see Chapter 4).

2.4.3 Summary

With regard to static image viewing, the evidence presented hitherto suggests the following course of events. During the first fixation on an image, the spatial layout of salient regions is acquired by bottom-up processes, based on visual saliency in different feature channels.

Depending on image nature, this information might be supplemented by an appropriate scene schema. Based on this information, the most salient spots are now foveated in turn. An inhibition of return mechanism prevents the eyes from switching back and forth between the two most salient spots. In addition to the saliency map, a map of importance or informativeness is built up concurrently during the first few fixations, becoming more and more detailed over time. With the gradual build-up of a representation of the image, the strategic, top-down processes can take over and exert their influence on the viewing patterns. As Antes (1974) observed, the most informative locations seem to be a kind of "operational basis" for visual exploration. The gaze often returns there for a short fixation before jumping to the next area. If a fixated location and its surroundings turn out to contain informative objects, they are likely to be examined in more detail immediately by few short saccades or re-fixated again later on. Moreover, task constraints further bias the selection of foveation targets, and interruptions of various kinds are possible. For instance, the viewer might realise that something "seems to be wrong" with the acquired data, like when the global meaning is unclear (Rayner & Pollatsek, 1992), resulting in an appropriate adjustment of scanning behaviour.

But it has to be stressed that it is improbable that a sharp demarcation between something like an initial "survey scan" and a later "detailed looking" phase (Buswell, 1935) can be found. Rather, a continuous shift of preponderance from bottom-up to top-down processes – which most likely operate in parallel – should be imagined. This notion receives strong support from several publications by Mannan *et al.* (Mannan et al., 1995; Mannan, Ruddock, & Wooding, 1996; Mannan et al., 1997b; summarised in Mannan, Ruddock, & Wooding, 1997a), based on a single set of data. They presented a series of 18 black-and-white images of real-world scenes in different versions, unfiltered, low-pass filtered and high-pass filtered. They found a high congruence of fixated locations between participants during the first 1.5 s, with re-fixations as a significant factor causing these high similarity values. This is consistent with the idea that initial behaviour is guided by bottom-up processes which are not expected to be strongly idiosyncratic. Moreover they found a high similarity within participants across the three different image versions. It is especially interesting that the low-pass filtered images elicited the same behaviour, as details which would be necessary in order to extract semantic information are significantly reduced. As for the bottom-up aspects, the authors found that only edge density and image contrast distribution was significantly correlated with fixation distribution, but not extreme values of luminance, or high spatial frequencies. They conclude "for brief presentations, eye movements made during examination of an unfamiliar image are performed automatically in response to the spatial features of the image" (Mannan et al., 1995, p. 363).

3 Experiment 1: Image size and saccade amplitude

As I have summarised in the previous chapters, there are many factors whose potential to affect eye movements during scene perception has not been scrutinised yet. One of these very basic variables is the size of the images as they are presented to the participants of an eye movement study. Display size is one of the first questions one has to decide on when setting up any image perception experiment, thus it seems important to know the possible consequences of this decision.

3.1 Theoretical part

Most scene perception studies hitherto have focussed on the characteristics of fixations, as they are thought to be closely related to ongoing perceptual and conceptual processing. The properties of saccades during scene image viewing remain a relatively understudied issue. In this context, the most important saccade property is saccade amplitude, characterised – for instance – by overall mean or median values, or by analysing the change of saccade amplitudes over viewing time.

In general, it is not more difficult for the oculomotor system to program and execute long saccades compared to shorter ones. Due to the minimum pause time between two saccades (as postulated by Salthouse & Ellis, 1980), and the insight that the time needed to program a saccade does not greatly vary for different saccade amplitudes (Irwin et al 2000; Loftus, 1981; Salthouse and Ellis, 1980), it would be uneconomical to move the eye with several small saccades instead of a single long one in order to foveate a target (see also Chapter 2.1.4). This logically leads to the hypothesis that saccade amplitude crucially depends on the spacing of the image elements that are important for the task to be solved. Thus, if an image is scaled to different sizes, saccade amplitudes can be expected to vary proportionally. This is also suggested by the fact that in reading studies, over a large range, saccade size is constant if measured in number of character spaces, not visual angle (Legge et al, 1985; Morrison and Rayner, 1981; for a review see Rayner, 1998). Thus it is a customary practice to indicate saccade amplitude in text characters, i.e. relative to letter size and spacing, and not in degrees of visual angle.

The hypothesis that saccade amplitude proportionally varies with image size might seem trivial. However, a recent comprehensive review (Henderson and Hollingworth, 1998) concluded that for scenes between 10 and 20°, mean saccade amplitude ranges between 2 and 4° *irrespective of scene size*. Similarly, Rayner and Pollatsek (1992) estimated an average saccade length of 4° for scene viewing *in general*. These statements, however, were descriptive in nature. Neither of these reviews indicated a theoretical reason why this should be so, nor substantiated this contention with new empirical data.

To the best of my knowledge, the only study designed to directly examine saccade amplitudes as a function of stimulus size was undertaken by Enoch (1959). In the context of a series of studies aimed at investigating natural search tendencies, a series of experimental aerial maps were presented to twelve participants, who were asked to search for a specific critical detail. The nine different maps were shown in grey-scale, full size was 9 x 9 in. The size of the display was varied by placing paper masks with circular apertures with different diameters in front of the map. This procedure resulted in circular displays of 3°, 6°, 9°, 18°, and 24° (diameter) plus the full square map which measured 51° 18' (diagonal). Of the several variables that were acquired, "average interfixation distance" (i.e. mean saccade amplitude) is of particular interest for the current topic. Enoch found that mean saccade amplitude almost linearly increased with the size of the display. Together with the finding that fixation duration decreased with increasing display size, this was interpreted as "greater concentration upon detail in the smaller displays" (Enoch, 1959 p. 285). It has to be pointed out that the size manipulation method used by Enoch left the spacing of image elements constant. Thus, whether this result would also occur when images are *scaled* to different sizes, thereby changing object spacing as well, has yet to be demonstrated.

Apart from Enoch's early report, there is a considerable number of image viewing studies documenting saccade amplitudes, but only on a single image size[4]. However, there are several reasons why these studies seem difficult to compare:

(1) A wide range of stimuli and tasks have been applied, such as line drawings in free viewing (Loftus and Mackworth, 1978) or memory (Henderson and Hollingworth, 1998) tasks, photographs of real-life scenes in a memory task (e.g. Mannan et al, 1995), or preference ratings of shaded drawings (Antes, 1974). Stimuli have been displayed in colour or grey-scale, and in different sizes, from 10.4° x 10.4° (Mannan et al, 1995) to 60 x 42° (Andrews & Coppola, 1999) or even the full visual field during everyday activities (Land & Hayhoe, 2001; Land, Mennie, & Rusted, 1999) or indoor scene perception (Tatler, Gilchrist, & Land, 2005).

(2) Saccade amplitudes have been specified by diverse distribution measures such as means, modes, or medians, or by analysing their temporal course.

(3) The obtained saccade range is crucially dependent on the tracking apparatus as well as data parsing methods and parameters. In particular, the ability to record small saccades considerably differs, which greatly affects the distribution measures of the data. Recent eye tracking systems operate with sampling rates up to 1 kHz, high spatial accuracy, and configurable, automatic parsing algorithms to determine fixations and saccades. As a contrasting example, Antes (1974) recorded corneal reflections with eight frames per second, and the resulting film was "manually" rated on a frame-to-frame basis. With this method, it is possible that – in particular – small saccades were overlooked.

Apart from global distribution measures such as mean and median values, the temporal course of saccade amplitudes during the viewing period is also of interest. To my knowledge, the only contribution to this topic was reported by Antes (1974). Based on earlier studies (Buswell, 1935; Karpov et al, 1968), he investigated the theory that image viewing behaviour is divided into an early "survey scan", characterised by short fixations and long saccades, and a later "detailed looking" phase with the opposite pattern. He presented ten shaded drawings to twenty participants, who were asked to give a preference rating. The stimuli were nine shaded drawings of natural scenes, taken from the Thematic Apperception Test (TAT), and a reproduction of *Morning on the Cape* by Leon Kroll. The images subtended less than 20° horizontally and vertically, and were shown for 20 s apiece. The eye movement recording system was a corneal reflection system that recorded the reflection image on 16 mm film with 8 frames per second; fixation data were scored by a frame-by-frame analysis of the film. The experiment indeed revealed that saccades during the hypothesised survey scan were longer than later on during the trial (see Fig. 5). The data also suggest that there is no sharp demarcation between the two postulated phases.

---

[4] see Chapter 3.4 for a more detailed account of these studies.

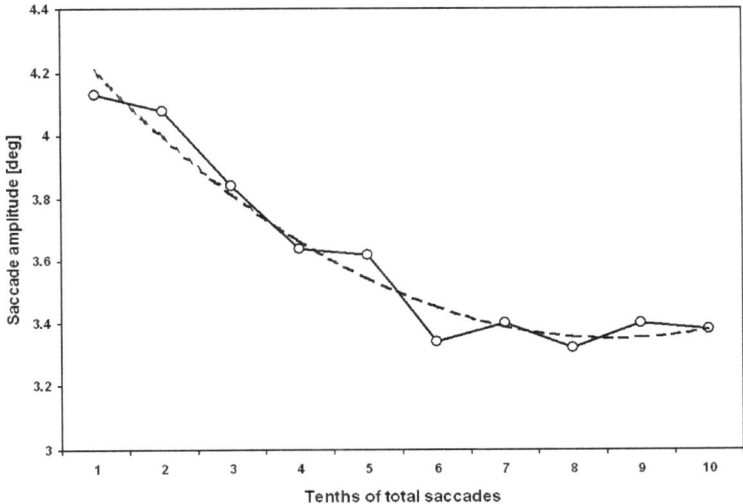

Fig. 5: Results of Antes' study (1974), redrawn after Fig. 3 in the original article. Viewing time was 20 s, thus one time window equals 2 s. The dashed line represents a 2nd order polynomial trendline.

However, a tentative analysis of data from another study conducted in our lab (von Wartburg et al., 2005) yielded a slightly different picture. In that study, complex images were presented in colour and grey-scale versions. The data from the colour images were re-analysed in the temporal domain. In Fig. 6, the results are plotted in five time windows of one second each. This analysis suggests that saccade amplitudes might initially increase and decrease only after that. However, it is possible that the initial low saccade amplitude values were due to the fact that, before each image, a central fixation stimulus had been shown. Therefore, starting position was constrained to the screen centre, which might in turn lead to shorter initial saccades as "important" objects are – in common photographs – often located near the middle of an image.

— Experiment 1: Image size and saccade amplitude

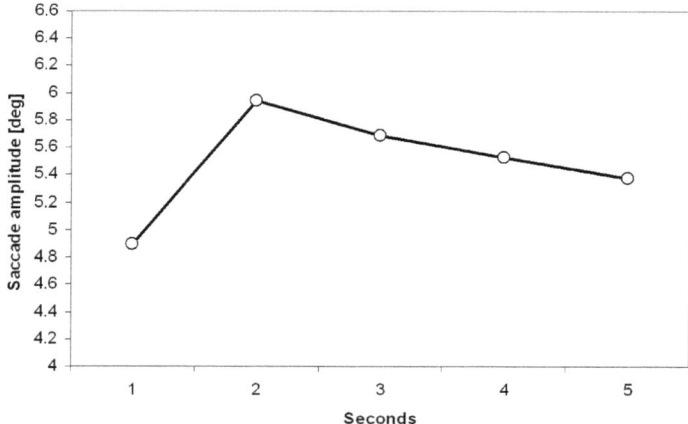

Fig. 6: Analysis of the temporal course of saccade amplitude change, based on data from an earlier experiment (von Wartburg et al., 2005). The graph represents mean saccade amplitude in five time windows of one second each.

In sum, these results suggest that this topic is in need of further investigation. The present experiment was intended to contribute to a better understanding of this issue. From a theoretical viewpoint, initial shorter saccades would make sense. When a novel image is presented to an observer, it might be efficient to first fixate image elements in the near vicinity of the starting point before attending to more distant objects. Thus I hypothesised that, even if there is no central fixation stimulus constraining starting position to the screen centre, initial saccades are shorter.

3.2 Methods

3.2.1 Participants

Forty-eight participants aged between 19 and 32 years ($M = 23.02$, $SD = 4.50$, 42 women and 6 men), mainly undergraduate students, volunteered for the study. No financial or other reward had been offered for their participation. All had normal or corrected-to-normal visual acuity, no strabismus, and normal colour vision. The latter was assessed by means of Ishihara's Test for Color Deficiency (Ishihara, 1999), while acuity was verified in a pragmatic manner after fitting the eye tracking equipment, by presenting a display with text in different sizes, and verifying that the participants were able to read the smallest line of print. All participants were naïve as to the experimental hypotheses. The experiment was undertaken with the understanding and written consent of each participant. This work adheres to the Code of Ethics of the World Medical Association (Declaration of Helsinki), as well as to the ethical guidelines defined by the Swiss Academy of Medical Sciences SAMS („Integrity in Science: Guidelines of the SAMS for scientific integrity in medical and biomedical research and for the procedure to be followed in cases of misconduct", 2002). The study was formally approved by the local ethics committee (Ethisches Komitee der Universität Bern).

3.2.2 Stimuli

The stimulus images were colour photographs of real-life scenes, scanned from illustrated books and magazines. As scenes, I understand "specific views of the environment within which we are embedded" as recently suggested by Henderson and Ferreira (2004). Therefore only such images were chosen that had been photographed from everyday vantage points and with approximately normal focal length, i.e. no aerial views, wide angle images or otherwise distorted perspectives. The images comprised landscape scenes, images of buildings, and populated city scenes (for examples see Fig. 7 top row, and Appendix A).

— Experiment 1: Image size and saccade amplitude

| Size 1: 10 x 7.7° | Size 2: 18 x 13.8° | Size 3: 26 x 19.8° | Size 4: 34 x 26° |

Fig. 7: Top row: Examples of stimulus images. For the complete set of images see Appendix A. Bottom row: Example image in the four different sizes. In the experiment, the images were presented in full colour.

Each full resolution original image was scaled down to four different sizes, in a way that maximal extent (i.e. image width) increased in a linear fashion. The resulting image sizes were 10 x 7.7°, 18 x 13.8°, 26 x 19.8°, and 34 x 26° (full screen). The size-reduced images were shown centred on the computer display, with borders in 50 % grey (Fig. 7 bottom row). In the following, image size will be numbered from 1 (smallest) to 4 (largest).

3.2.3 Apparatus

The images were presented in a dimly lit room on a 21" LCD flat panel display (Samsung SyncMaster 213T) with a resolution of 1600 x 1200 and 16 bit colour depth, driven by an nVidia GeForce MX440 graphics adapter. Active screen size was 43.2 × 32.3 cm and viewing distance 70.5 cm, resulting in a maximal viewing angle of $34 \times 26°$.

Eye position was recorded with an infrared video-based tracking system (EyeLink[TM], SensoMotoric Instruments GmbH, Teltow, Germany). This system consists of a headset with a pair of infrared cameras and illuminators recording the eyes, and a third camera monitoring screen position in order to compensate for any head movements. Thanks to this compensation mechanism, no head clamp, bite bar or other head fixation was necessary, which made the procedure less unpleasant for the participants. Only a chin rest was used to reduce head movements and ensure correct height and distance of the eyes relative to the stimulus display. The system provides a temporal resolution of 250 Hz, a net resolution of 0.01° (noise limited) and a gaze-position accuracy relative to the stimulus position of 0.5° - 1.0°, largely dependent on participants' fixation accuracy during calibration. The minimal differences between the movements of both eyes that can occur are negligible if the participants have no strabismus, and therefore gaze position was derived from one eye only. Each image block was preceded by a 3 × 3 point grid calibration scheme which is part of the Eye-Link system and allows for the detection and correction of non-linearities, as well as correction of headset position changes between blocks. In addition to the calibration procedure incorporated in the EyeLink system, a second 3 × 3 point grid calibration sequence preceded each stimulus block. The system was supplemented by an off-line computer-interactive calibration and data evaluation software developed in our laboratory (running under Matlab 6.1 Release 12.1, The MathWorks, Inc.), which made use of the data from the second calibration sequence allowing for an additional improvement of calibration. This results in a saccade amplitude accuracy of typically better than 0.5° over a trial block of 90 s duration.

Eye monitoring was conducted on-line throughout the image blocks. The EyeLink system parsed the raw eye tracking data for saccades in real time. Three thresholds were used for saccade detection: Movement distance, velocity, and acceleration. An eye movement was considered a saccade when either velocity exceeded 35°/s or acceleration exceeded 9500°/s^2, and a movement distance of more than 0.1° was measured. This parameter set is tuned to use for cognitive research, as recommended by the manufacturer of the system. In our lab, these settings have proven to be useful during the last five years and have been used for more than a thousand participants in different studies.

3.2.4 Experimental procedure

The images were presented in twelve blocks of eight images. As described above, every block was preceded by a calibration sequence, which terminated with a central fixation dot. Thus, the starting point for the first image would have been constrained to the screen centre, which in turn might have created a bias for the first saccade to be shorter. To prevent this, an additional image – a similar scene – was shown in full screen size as the first stimulus of every block, but was excluded from all analyses. Viewing time was 5.5 seconds per image. In order to provide participants with a well-defined task, they were instructed to view the images in preparation for a recognition test to be carried out immediately after the last image of each block. This test consisted of a display with eight image pairs; in each pair, one image had been presented before, the other one was a new image. An example of a recognition display is shown in Fig. 8. It has to be pointed out that the recognition test was only included to provide the participants with a well-defined task, and not to acquire data for any kind of performance evaluation.

Each block consisted of images that were thematically similar, e.g. city, mountain, rural, or seaside scenes, in order to make the recognition task more plausible (see Appendix A). Image presentation sequence was organised as follows. Every participant viewed each of the 96 different images once, one fourth in size 1, one fourth in size 2 etc. In every block, two images in each of the four sizes were shown. This was balanced in a way that in the end, every image had been viewed by twelve participants in every size. The order of images and the sequence of sizes within a block were fully randomised for all participants, as well as block sequence.

Fig. 8: Example recognition display. In the experiment, the images were presented in full colour.

Each participant served for an experimental session of one hour at most, including instruction and debriefing; the time in the eye tracker never exceeded half an hour. In the instruction phase, the participants were shown an example image block and a sample recognition display, so they had the opportunity to make themselves familiar with the type of images they were to expect, the presentation duration, and the recognition task.

3.2.5 Data analysis

To account for the fact that two consecutive images in randomised size followed each other immediately (i.e. without any other frame in-between), those saccades a participant performed to "jump into the image area" in cases when a smaller image followed a larger one were excluded from the analysis. For the descriptive analysis of saccade amplitude distribution, saccades were pooled over all participants and images, separately for the four image sizes. Based on these data, the descriptive values, i.e. mean, median and mode were computed. Moreover, a histogram analysis was undertaken to illustrate the overall distribution of saccade amplitudes. Histogram calculations were performed with a bin width of $0.5°$.

For inference statistics, two measures were analysed: As saccade amplitudes are not normally distributed, first the median and modal value of saccade amplitude per participant and image size were calculated over all images. Moreover, mean values were obtained as well, so as to be able to compare the results of the present study with earlier publications that solely report mean saccade amplitude (see Chapter 3.4).

For the analysis of the temporal course of saccade amplitudes, the mean amplitude of all saccades made by the participant on the current image was subtracted from each measured saccade. Then, these values were pooled over all images, in five time windows of one second duration each, the first window starting at the outset of the first valid saccade. In this way, the temporal variation of amplitudes irrespective of the overall mean values was obtained.

For all inference statistics, general linear model (GLM, McCullagh & Nedler, 1989) repeated measures statistics as implemented in SPSS 12.0.1 (SPSS Inc., Chicago IL) were used. Deviations from sphericity[5] were tested by Mauchly's Test of Sphericity and degrees

---

[5] The spericity assumption means that the variances (pooled within-group) and covariances (across participants) of the different repeated measures are homogeneous (identical).

of freedom corrected by a correction factor (method of Huynh-Feldt) if this assumption was violated. For post-hoc testing, multiple pairwise comparisons were applied.

Even though the distribution of saccade amplitude is not normally distributed, it is safe to assume that the calculated saccade *mean* and *median* values may be considered as being sampled from a normal population. In order to verify this normality assumption, all analysed variables as well as their standardised residuals after the GLM statistics were subjected to statistical normality tests. Neither the Kolmogorov-Smirnov nor the Shapiro-Wilk test of normality found any significant deviation from normality. Furthermore, GLM statistics are remarkably robust against violations of the normality assumption (for a summary, see Lindman, 1974).

As an aid for the interpretation of the quantitative results (Fig. 14), a graphical representation method for spatial fixation distributions was devised. The basic idea is to calculate a fixation density map in the first step, and to visualise it by means of a so-called contour plot (Matlab 6.1 Release 12.1, The MathWorks, Inc.). The fixation density map was calculated as follows: On a virtual image area with the same dimensions as the original image (i.e. 1600 × 1200 pixels for the current experiment), for every fixation, a three-dimensional Gaussian "bell shape" is set onto the area (see equation below), thus resulting in a "mountain landscape" of superimposed Gaussian "bumps".

$$z = \{\alpha \cdot t + (1-\alpha)\} \cdot e^{\sigma^2}$$

x, y	screen co-ordinates of fixation in pixels
σ	"width" of gaussian bell shape (e.g. hypothetical fovea diameter)
α	contribution of fixation duration. Range from 0 (duration disregarded) to 1 (full effect)

The resulting data matrix is then subjected to a contour plot algorithm which draws "equidistant" lines onto the viewed image.

3.3 Results

Fig. 9 shows the distribution histograms of saccade amplitudes pooled over participants and images, and Table 1 lists the distribution measures. For a detailed results table of histogram data see Appendix B, Table B-1.

While the modal value of all four data sets was identical, the other distribution measures substantially differed. Concerning the skewness measure, a positive skewness value of about 1 was found for all sizes, i.e. an asymmetric distribution with a long right tail. Kurtosis, on the other hand, was highest for small images, low for size 2 and 3, and somewhere in-between for size 4.

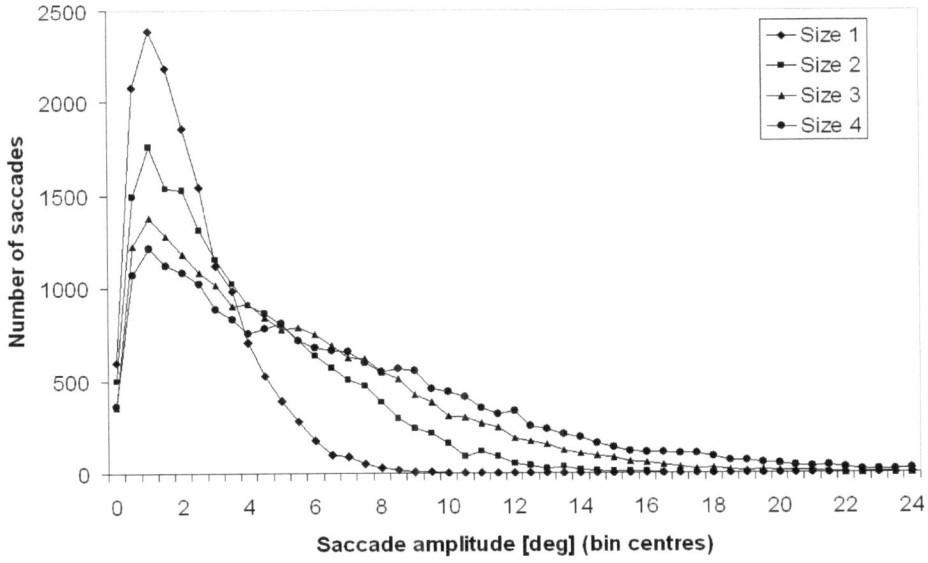

Fig. 9: Histogram plot of saccade amplitude distribution. Bin width for histogram calculation was 0.5°.

— Experiment 1: Image size and saccade amplitude

Table 1: Distribution measures for pooled saccade amplitudes.

	Size 1	Size 2	Size 3	Size 4
number of saccades	14808	17281	18468	19359
mean [deg]	2.218	3.764	5.131	6.297
median [deg]	1.867	3.068	4.293	5.204
mode [deg]	0.9	0.9	0.9	0.9
standard deviation [deg]	1.554	2.794	3.873	4.923
skewness	1.164	0.959	1.019	1.127
standard error of skewness	0.020	0.019	0.018	0.018
kurtosis	1.754	0.482	0.871	1.269
standard error of kurtosis	0.040	0.037	0.036	0.035

However, the range of very short saccades (i.e. < 1°) is problematic for the parsing algorithms used to extract saccades from raw eye movement data. In consequence, the modal value of about 0.9° found for all four image sizes might be a measurement artefact. Therefore, the raw data from all four sizes were pooled and re-parsed with different saccade detection thresholds, i.e. with modified minimal speed and minimal acceleration values. The detailed results are given in Appendix B, Table B-2. The results plotted in Fig. 10 show that with lower thresholds, the modal values also shift towards lower values. At the extreme, the distribution tends to extend into the range of micro-saccades.

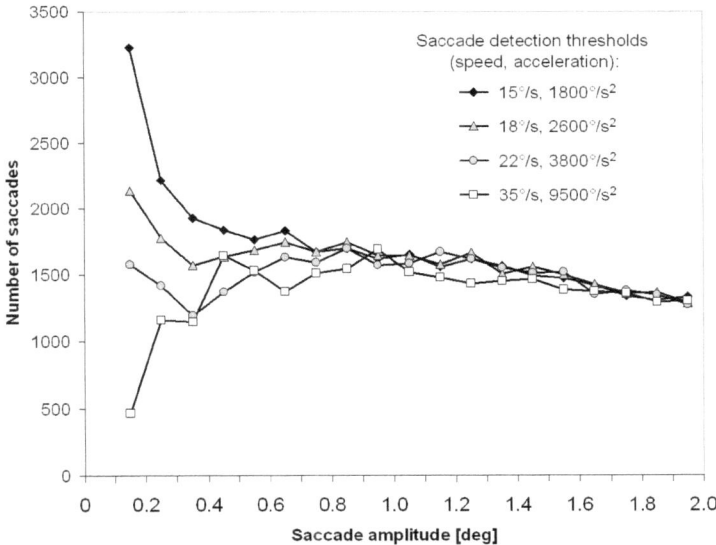

Fig. 10: Comparison of low amplitude saccade distribution, parsed with different settings. The algorithm detects a saccade if the eye moves at least 0.1° and either the speed of the movement is greater than the speed threshold or its acceleration exceeds the acceleration threshold. The histogram was calculated with bin width of 0.1°. In the first analysis, the parameters were 35°/s and 9500°/s^2.

In sum, when looking at the range from 1° to maximal image extent exclusively, the peak of saccades is found at the low end of the distribution, with a continuous decrease towards longer saccades.

In the range from 1° and approx. 75% of image extent, the resulting distribution can be reasonably well approximated by a quadratic or cubic curve (see Table 2 and Fig. 11).

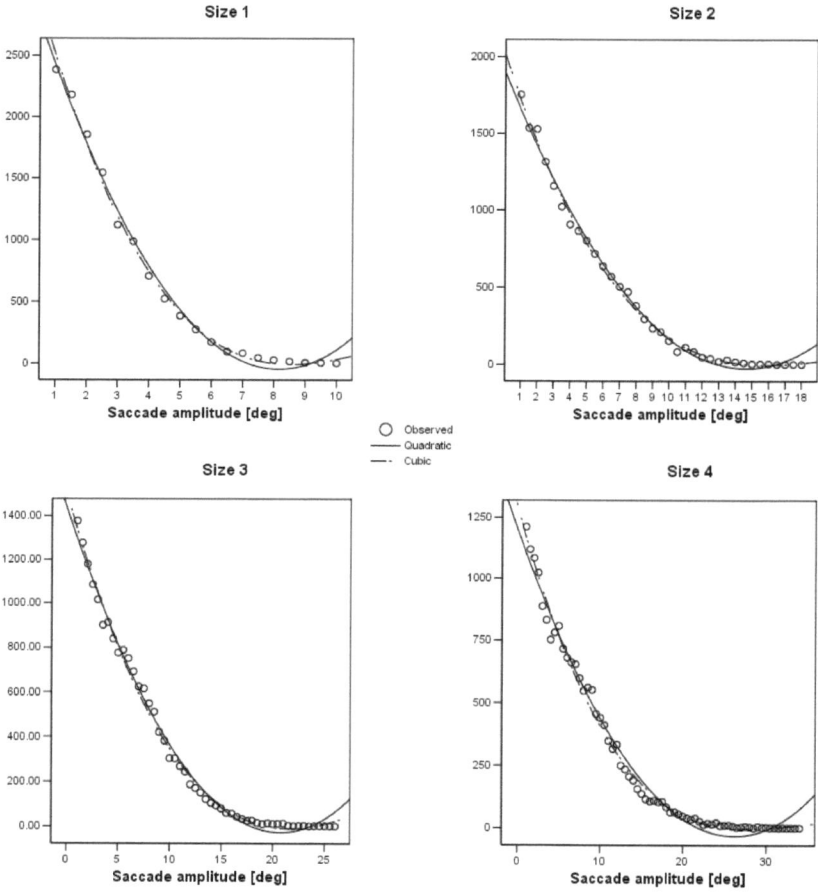

Fig. 11: Saccade distributions between 1° and maximal image extent, with fitted quadratic and cubic curves.

Table 2: Curve fitting results

		df	F	Sig	b0	b1	b3	b3
Size 1	quadratic	16	1137	<.001	3160	-786.1	48.1	
	cubic	15	1200	<.001	3400	-987.0	91.0	-2.600
Size 2	quadratic	32	2441	<.001	1915	-264.8	9.024	
	cubic	31	3036	<.001	2036	-329.1	17.18	-0.286
Size 3	quadratic	48	3444	<.001	1435	-141.1	3.399	
	cubic	47	3160	<.001	1501	-166.7	5.708	-0.057
Size 4	quadratic	64	2344	<.001	1199	-94.08	1.796	
	cubic	63	2865	<.001	1290	-122.3	3.773	-0.038

Mean and median saccade amplitude values are presented in Table 3 and plotted in Fig. 12. Repeated measures GLM statistics yielded a significant main effect of image size for both measures (mean: $F = 1695$, $df = 1.763$, $p < .001$; median: $F = 719$, $df = 1.991$, $p < .001$). All pairwise comparisons were significant with $p < .001$. The correlations of image size with the overall mean and median values were highly significant (mean values: $r = .998$, $p = .002$; median values: $r = .998$, $p = .002$); the curves are best approximated by linear equations, with the coefficients $y = 0.606 + 0.170x$ for the mean curve, and $y = 0.508 + 0.143x$ for the median curve, where x represents image size (i.e. maximal image extent).

Table 3: Mean, median, and relative saccade amplitude values.

		Size 1	Size 2	Size 3	Size 4
Mean amplitude [deg]	M	2.209	3.765	5.132	6.290
	SD	0.327	0.470	0.667	0.787
Median amplitude [deg]	M	1.871	3.115	4.355	5.269
	SD	0.324	0.529	0.773	0.866
Relative amplitude [%]	M	18.71	17.31	16.75	15.50
	SD	3.241	2.940	2.974	2.552

— Experiment 1: Image size and saccade amplitude —

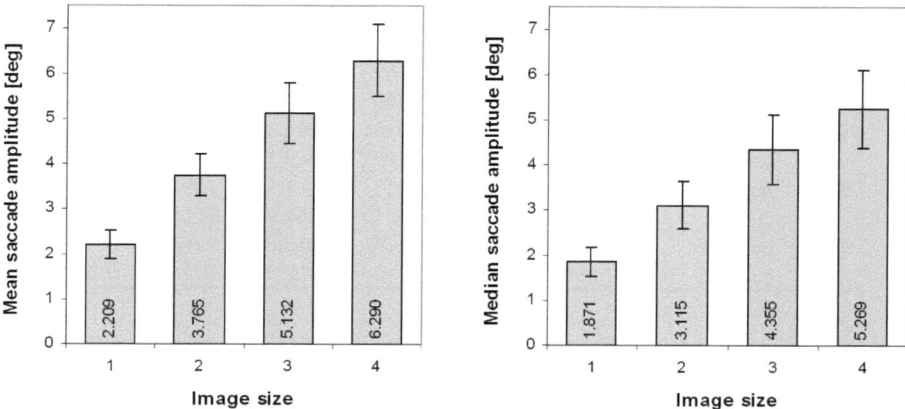

Fig. 12: Saccade amplitude graphs. Left panel: Mean amplitude. Right panel: Median amplitude. Error bars indicate the standard deviation. The main effect of size is highly significant ($p < .001$) for both mean and median, as well as all pairwise comparisons.

These results strongly support my hypothesis that saccade amplitude is directly proportional to image size when image content is held constant, and thus contradict the view that saccade amplitude is independent of image size.

This leads to the question whether saccade amplitude might be identical when calculated relative to image size, which is suggested by the fact that, in reading studies, saccade size is constant if measured in number of character spaces, not visual angle. Thus, it might also be the case in scene image perception. To this end, saccade amplitude was computed in per cent of maximal image extent, i.e. image width. The results of this calculation are given in Table 3 and Fig. 13. The plot indicates that saccade size is not constant if measured in per cent of total image extent. Rather, relative saccades are longest on small images, and shortest on large images.

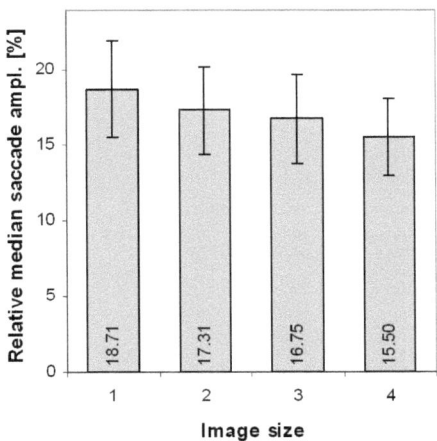

Fig. 13: Median amplitude relative to image size. Error bars indicate the standard deviation.

— Experiment 1: Image size and saccade amplitude

To see whether this might (partly) be caused by saccades transgressing the actual image area, fixations falling outside this area were counted. However, the percentages of fixations missing the image were rather low, with 1.105 % for size 1, 0.458 % for size 2, 0.276 % for size 3, and 0.201 % for size 4, thus I conclude that this is not the case. For a graphical representation of fixation distribution over all images and participants see Fig. 14.

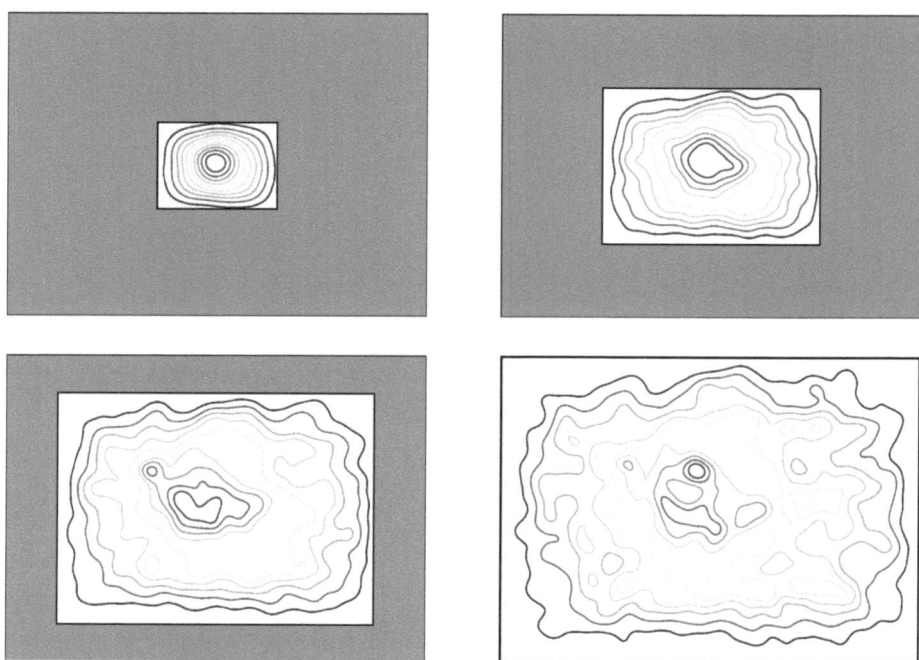

Fig. 14: Fixation density plots over all images and participants, separately for the four sizes. The highest density of fixations is found in the central regions.

The results of the temporal analysis of saccade amplitude are plotted in Fig. 15 and given in Appendix B (Table B-3). Note that these values are not absolute amplitudes, but represent the difference value of each saccade minus the mean amplitude of all saccades made by a certain participant on the current image. Repeated GLM statistics confirmed a main effect of time ($F = 11.52$, $df = 3.255$, $p < .001$), whereas image size had no significant effect on the difference values ($F < 1$). The results of the post-hoc tests for the time main effect were $p < .001$ for 1st vs. 2nd window, $p < .01$ for 2nd vs. 3rd window, $p = .054$ for 3rd vs. 4th window, and $p < .05$ for 4th vs. 5th window.

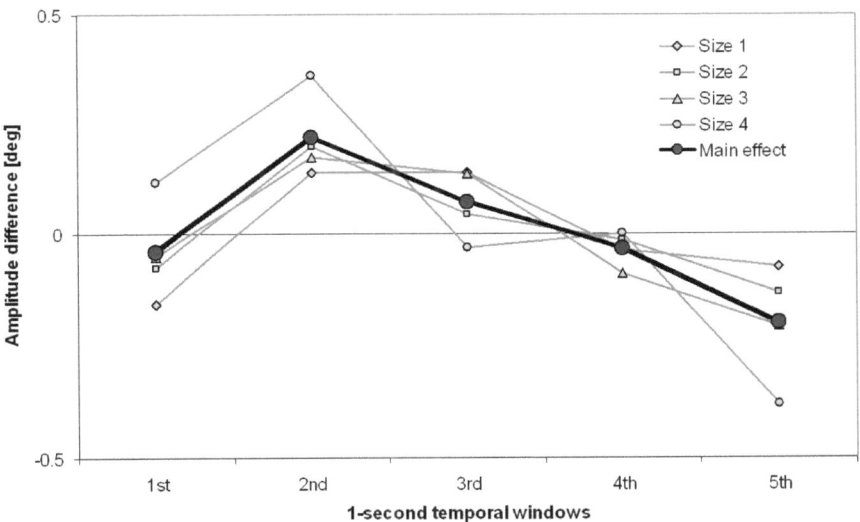

Fig. 15: Temporal course of saccade amplitude change. Evaluation for five one-second time windows. The values represent the temporal variation of amplitudes irrespective of the overall mean values. The bold line indicates the main effect of time, i.e. the mean values over all four image sizes.

The results show a clear tendency for the saccade amplitudes in the first time window to be shorter, reach a maximum during the second time window, and slowly decrease afterwards. On the one hand, this confirms my hypothesis that initial saccades are shorter during scene perception tasks (cf. Fig. 6). On the other hand, the current results also confirm Antes' findings (1974) that overall, saccade amplitudes slowly decrease (see Fig. 5).

3.4 Meta analysis of earlier studies

As a second step, I wanted to investigate whether the results from the present experiment fit into results from other studies. Several databases (PubMed, PSYNDEXplus, PsycINFO) were searched for articles investigating eye movements during complex image viewing. The following selection criteria were applied: (1) Only studies using natural scene images were included, independently of whether they were drawings, black-and-white or colour photographs. (2) Any saccade amplitude measure specifying amplitude distribution over the complete viewing time had to be available in the report. If the values were not listed as means, the mean values for the current meta analysis were estimated from what was given. (3) For fixed-head setups, only studies presenting the stimuli smaller than 40° (horizontally) were included. In my opinion, using images larger than approx. 35° is not admissible when head movements are restricted, for it is unnatural to make such large saccades without concomitant head (or body) movements.

It has to be pointed out that none of these studies was explicitly aimed at studying saccade amplitudes in relation to image size. Consequentially, the stimuli were displayed in one size only. Following those criteria, the search resulted in ten relevant studies, which are briefly described below in alphabetical order.

Antes (1974): The first paper investigates cognitive determinants of fixation location during picture viewing. Stimuli were nine shaded drawings of natural scenes, taken from the Thematic Apperception Test (TAT), and a reproduction of *Morning on the Cape* by Leon Kroll. All images were prepared as achromatic slides and were of approximately the same size, subtending no more than 20° horizontally and vertically. Twenty participants were shown the images for 20 s each and were asked to decide which images they preferred. The eye movement recording system was a corneal reflection system that recorded the reflection image on 16 mm film with 8 frames per second; fixation data were scored by a frame-by-frame analysis of the film. For the current analysis, mean saccade amplitude was calculated based on the data plotted in Fig. 3 in Antes' article.

van Diepen et al. (1995): This paper reports a study into the chronometry of foveal information extraction during scene perception. Eight participants explored 31 different line drawings of realistic scenes with the task to find non-objects in the scene. Target objects were embedded in twelve of the images, at various positions. At each fixation, foveal information was masked after a brief interval. Stimuli subtended approx. 16 x 12°. Saccades were measured with a Generation-V dual-Purkinje-image eye tracker with 1000 Hz sampling rate. Only the mean saccade amplitude value from the "No mask" condition entered the current meta analysis (see Fig. 2 right panel in the article).

Henderson and Hollingworth (1998): In this overview, the authors also report a study aimed at the comparison of viewing behaviour in response to colour photographs of real-world scenes, line drawings of the same scenes, and computer-rendered colour images of similar scenes. The images subtended 14.5 x 10°. Eight participants examined each of 30 scenes for 15 s each. They had been told that afterwards they would be given a memory test in which they would have to discriminate the test scenes from new scenes in which only a small detail of a single object might be changed. A dual-Purkinje image eye tracker with a sampling rate of 1 kHz was used. Mean saccade amplitude was 2.4°, independently of image type.

Henderson et al. (1999): In this research project, the authors investigated the effects of semantic consistency on eye movements during complex scene viewing. The same experimental setup as in the above cited paper was used, i.e. image size of 14.5 x 10° and eye tracking with a 1 kHz dual-Purkinje tracker. Eighteen participants viewed 24 line drawings of everyday scenes, partly containing contextually inconsistent items (Experiment 1 reported in the paper). They were instructed to expect a memory test afterwards, in which they would have to discriminate the test scenes from slightly modified new scenes. A value of 3.03° was indicated as average saccade amplitude.

Loftus and Mackworth (1978): This paper describes an investigation of the cognitive determinants of fixations location during picture viewing. Each participant was shown 78 line drawings at 4 s per image; some of the images contained "informative" objects, e.g. an unexpected object such as an octopus in a farm scene; images subtended approx. 30 x 20°. Participants were instructed to expect a recognition test afterwards. Only 31 of the 78 image groups were selected for analysis. Eye movements were recorded by a digital, pupillary-reflection camera that recorded eye position on a 18-frame/s movie. Eye position was detected based on a 8 x 5 grid laid over the pupil reflection image. In the article, average interfixation distance is indicated as being in the order of 6.5 to 8°, thus an estimated value of 7.25° entered the meta analysis.

Mannan et al. (1995): An experiment aimed at investigating the automatic control of saccadic eye movements made in visual inspection of briefly presented 2-D images is reported. Eleven black and white photographical images of natural scenes were presented for three seconds, in three versions, unfiltered, low-pass and high-pass filtered. Image size was 10.4 x 10.4°. Eighteen participants participated in the study; they had to view the images under the instruction to memorise them for unspecific questions that would be posed after the experiment. An infrared video-based tracking system (P-scan) with a temporal resolution of 50 Hz was used. Only the saccade amplitude average value over all images in the unfiltered condition entered the present analysis (cf. Fig. 11 in Mannan et al's article).

Parkhurst and Niebur (2003): By means of the reported experiment, the statistical properties of scene content selected by active vision were examined. Four paid participants freely examined 300 photographic scene images for five seconds each. 200 of the images were natural scenes (home interiors, natural landscapes and building and city scenes), the remaining 100 images were computer-generated fractals. The images were presented on a cathode ray tube (CRT) display and subtended 30 x 22.4°. Eye movements were detected by means of an infrared video-based tracking system (ISCAN model RK-416) with 60 Hz sampling rate. For the current analysis, mean saccade amplitudes from the three scene image categories (excluding the fractals) were averaged.

Saida and Ikeda (1979): This study was aimed at investigating the useful visual field size for pattern perception. Two experiments were performed, of which only the "large picture experiment" is relevant for this meta analysis. In this part of the study, five participants viewed 80 black and white drawings (out of 1600) of common scenes and objects each, for various exposure durations. Participants were instructed to view the images in preparation for a recognition task. Only one participant viewed the images in an unrestricted viewing condition, and only this data set is included in the present meta analysis. For the other participants, the images were degraded outside an eye-contingent window of specifiable size. The images were shown on a TV monitor with a size of 18.8 x 14.4°. Eye position was measured with a corneal reflection tracker, without sampling rate or saccade detection method being specified in detail. Mean saccade amplitude value had to be estimated based on a graph in the article (Fig. 5, top panel).

Shioiri and Ikeda (1989): Similarly as the above paper by Saida and Ikeda, this paper reports a study of the useful resolution for picture perception as a function of eccentricity. Three participants visually explored a large number (80 out of 3200) of black and white drawings of common scenes and objects on a 15 x 15° CRT display. Within an eye-contingent window of specifiable size, the image was shown in full resolution, whereas outside this window, resolution was degraded. Eye position was measured with a corneal reflection tracker; the sampling rate was not specified in the paper. The participants had to solve a recognition task after viewing the stimuli. For the current meta analysis, only the data from the condition with no degradation was considered. In the results graph, the authors reported a median saccade length of 3.2° in this condition. Based on this value, a mean value of 2.8° was estimated, which entered the present analysis.

von Wartburg et al. (2005): This study was designed to investigate the influence of colour on oculomotor behaviour during image viewing. Forty participants viewed 36 complex images in colour or grey-scale in preparation for a recognition task. Image size was 29 x 22°, eye movements were recorded with an infrared video-based tracking system (EyeLink™, SensoMotoric Instruments GmbH, Teltow, Germany; same equipment as used for this thesis). For the present purpose, only the data from the colour images were included.

The compiled mean saccade amplitude values are listed in Table 4, together with stimulus size, maximal extent (in either horizontal or vertical dimension), viewing task, and stimulus type. In addition, the results from Enoch's study (1959) are given, for it is the only study experimentally manipulating image size. However, it has to be noted that the images he used do not match the inclusion criteria listed above. For a detailed description of Enoch's study see Chapter 3.1. Moreover, the table is supplemented by the results from Experiment 1 of the present thesis.

— Experiment 1: Image size and saccade amplitude

Table 4: List of image viewing studies reporting saccade amplitudes, in alphabetical order. The table lists authors, mean amplitude, maximal extent of the images (in either horizontal or vertical dimension), the assigned viewing task, and a brief specification of the stimuli (b/w = black and white). Values are given in degrees visual angle. At the bottom of the table, the results from Enoch's study (1959) and the results from Experiment 1 are given.

Authors	Mean amplitude	Stimulus size	Maximal extent	Task	Stimuli
Antes (1974)	3.6	max. 20	20	preference rating	b/w shaded drawings from the Thematic Apperception Test
van Diepen, DeGraef, and d'Ydevalle (1995)	3.35	16 x 12	16	search for non-objects	b/w line drawings of real-world scenes
Henderson and Hollingworth (1998)	2.4	14.5 x 10	14.5	recognition	line drawings, photos, computer-rendered images of indoor scenes
Henderson, Weeks, and Hollingworth (1999)	3.03	14.5 x 10	14.5	recognition	b/w line drawings of real-world scenes
Loftus and Mackworth (1978)	7.25 [1]	30 x 20	30	recognition	b/w line drawings of real-world scenes
Mannan et al. (1995)	2.94	10.4 x 10.4	10.4	memorise for unspecific questions	b/w scene photos
Parkhurst and Niebur (2003)	6.74	30 x 22.4	30	free viewing	colour scene photos and fractals
Saida and Ikeda (1979)	3.5 [2]	14.4 x 18.8	18.8	free viewing, later recognition	b/w drawings of scenes and objects
Shioiri and Ikeda (1989)	2.8 [3]	15 x 15	15	recognition	b/w drawings of scenes and objects
von Wartburg et al. (2005)	5.69	29 x 22	29	recognition	colour scene photos, fractals, and abstract art paintings
Enoch (1959)	0.87	⌀ 3	3	search	b/w experimental aerial maps
	1.82	⌀ 6	6		
	2.13	⌀ 9	9		
	3.73	⌀ 18	18		
	4.33	⌀ 24	24		
	6.3	51.3 diagonal	36.3		
Experiment 1	1.871	10 x 7.7	10	recognition	colour scene photos
	3.115	18 x 13.8	18		
	4.355	26 x 19.8	26		
	5.269	34 x 26	34		

[1]) Estimated value. The authors reported an average distance on the order of 6.5°-8° of visual angle.
[2]) "Large picture experiment", estimated value based on Fig. 5 in the article, unrestricted visual field.
[3]) Estimated value. The authors reported a median saccade length of 3.2 deg.

A correlational analysis of mean saccade amplitude and maximal extent of the images yields a highly significant correlation (Spearman's $\rho = .959$; $p < .001$). That is, the correlation between image size and mean saccade amplitudes as reported by the studies included in this meta analysis accounts for 92 % of the variance.

In Fig. 16, mean saccade amplitudes of the studies included in the meta analysis are plotted as a function of the maximal extent of the stimuli, together with the results from Enoch's study (1959) and Experiment 1 of this thesis.

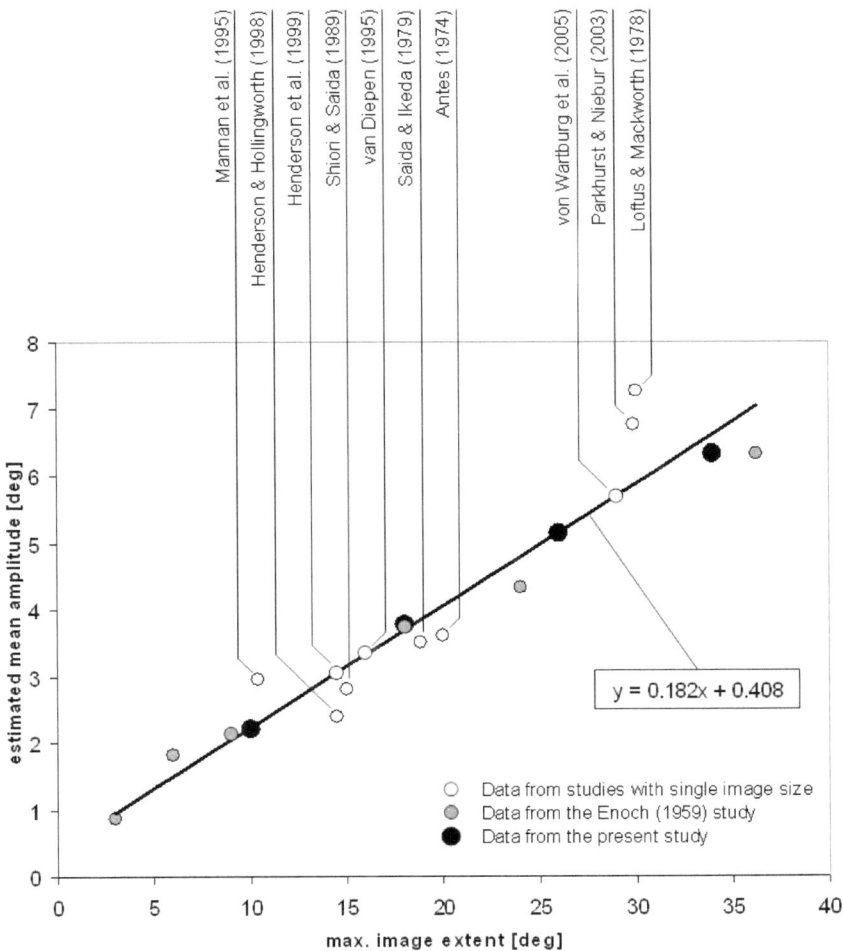

Fig. 16: Mean saccade amplitudes of the results reported in earlier publications, together with the data from the present experiment. Spearman's ρ = .959; p < .001.

Experiment 1: Image size and saccade amplitude

It is obvious that the results from the current experiment fit into the meta analysis results very well. Thus, overall mean saccade amplitude for any scene perception study can be approximated with reasonable accuracy by the following linear equation:

$$\text{mean saccade amplitude} = 0.408 + 0.182 \cdot \text{maximal image extent}$$

In other words, the present results confirm my hypothesis that mean saccade amplitude is directly proportional to image size. This is even so when data from a large range of studies, with differing stimuli, methods, or viewing tasks are considered.

3.5 Discussion

3.5.1 Overall saccade amplitude distribution

Experiment 1 was designed to investigate the relation between image size and saccade amplitude during scene perception. It revealed a nearly perfect positive correlation of mean as well as median saccade amplitude with image size. This contrasts with earlier review articles concluding that mean saccade amplitude is independent of image size (e.g. Henderson and Hollingworth, 1998; Rayner and Pollatsek, 1992). Now, an important question is whether this is an effect of image size as such, or an effect of object spacing, which was also affected by the image scaling procedure applied for the current experiment. Two arguments speak in favour of the view that image size is the more important factor. First, the results reported by Enoch (1959) – which is the only study that has directly addressed the question of how image size affects saccade amplitude hitherto – are very similar. In that experiment, however, image size was not manipulated by scaling, but by placing paper masks in front of the image. This procedure changes image size, but not object spacing. The high correspondence of Enoch's values with the current results is even more striking as his experimental conditions and eye tracking equipment were very dissimilar to the ones used in the present study. He used grey-scale experimental maps as stimuli, which were quite different from the colour scene images as they were used here, and his corneal reflection ophthalmograph certainly had a different ability to detect small saccades than the EyeLink equipment. The second argument for the view that image size is more important a factor than object spacing comes from the meta analysis of previous studies. It demonstrates that the size of stimulus images is *the* dominant factor influencing mean and median saccade amplitude during scene viewing. A clear linear relationship between image size and mean saccade amplitude was found, independent of factors such as image type and content, and therefore also independent of object spacing (cf. Fig. 16).

In consequence, mean and median saccade amplitude can reliably be predicted by means of a simple linear equation expressing saccade amplitude as a function of image size. This finding might suggest that it was reasonable to indicate saccade amplitudes in percent of image size, not degrees. In reading studies, this is a customary practice, as over a large range, saccade size is constant if measured in number of character spaces, not visual angle (Legge et al, 1985; Morrison and Rayner, 1981; for a review see Rayner, 1998). In other

— Experiment 1: Image size and saccade amplitude

words, if a text display is scaled to different sizes, saccade amplitudes relative to display size remain constant. However, this does not seem to be the case for saccade amplitudes in scene image perception. The analysis of median saccade amplitude relative to image size shows that the smaller a stimulus image is, the larger are the relative saccades. This might be an effect of the fact that, in an image that has been scaled down to a small size, more information is foveally available, and thus longer saccades can be made; or, in other words, less short, intra-object saccades are necessary. This possibility is illustrated in Fig. 17.

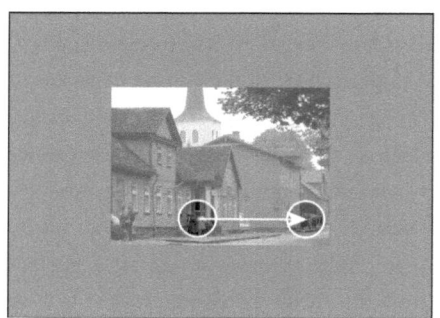 

Fig. 17: Illustration of a possible cause for longer relative saccades on smaller images. On the small image, to acquire approximately the same amount of information, only two fixations are necessary compared to the large image where three fixations are made. On the small image, the one saccade measures 51% of image width. On the large image, relative saccade amplitude is 30% (average of the two saccades, measuring 7% and 46%).

As to the overall saccade amplitude distribution, it might seem at first sight that there was an identical modal value at approx. 0.9° for all four image sizes. At a closer look, however, it becomes clear that this mode is an artefact of the saccade parsing algorithm. When its parameters are tuned for higher sensitivity for small saccades, the mode tends to shift towards shorter values. At the extreme, the range of "regular" saccades smoothly merges with the range of eye movement phenomena usually regarded as micro-saccades, micro-tremor and the like. That is, when only the range from approx. 1° to the maximal image extent is considered, the peak of saccades lies at the lower end of the distribution, with a continuously decreasing incidence towards longer saccades. The resulting distribution can be reasonably well approximated by a quadratic or cubic curve.

From a methodological standpoint, these findings underline that stimuli for scene viewing studies should be displayed in reasonable size (at least 20°) for two reasons. First, this improves the ratio between measured amplitude values and measurement error. Second, as I have demonstrated, it seems advisable to exclude saccades below approx. 1° as they are difficult to parse and differentiate from micro-movements that are not to be considered "regular" saccades. Using larger images minimises possible distortion caused by this filter procedure.

3.5.2 The temporal course of saccade amplitude change

Concerning the temporal course of saccade amplitude, two hypotheses were under consideration: (1) Saccade amplitudes steadily decrease with time (Antes, 1974). (2) Saccade amplitudes initially increase and then decrease (based on a re-analysis of results from von Wartburg et al., 2005). The current results strongly support the latter hypothesis. Saccades during the first second of viewing were found to be significantly shorter than those performed during the following second. After this peak, saccade amplitudes were slowly decreasing. This behaviour was found to be the same for all four image sizes. That is, independent of the overall saccade amplitude level, the same increase/decrease pattern was observed. Moreover, the present data demonstrate that the initial shorter saccades are not caused by a starting point of exploration being constrained to the screen centre, as it could have been reasoned on the basis of the above mentioned re-analysis.

This is another indication of the enormous speed and flexibility of the visuo-motor system: Saccades are scaled to the requirements of the current stimulus immediately, i.e. during the first fixation. An alternative hypothesis could have been to postulate that the visual system starts out with certain "default" saccade amplitude and only after that scales them to the actual image size. However, such a hypothesis is not supported by the present data.

The change of saccade amplitudes over time is relevant to a hypothesis dating back to the early reports by Buswell (1935) and Karpov et al (1968). They postulated that image viewing was divided into an early "survey scan" with long saccades (and short fixations), and a later "detailed looking" phase with shorter ones (and longer fixations). This was further investigated by Antes (1974). He indeed found that saccades were longest during the initial 2 s viewing phase, and slowly decreased over time. However, his data rather suggested the two phases to be the opposite ends of a continuum, and not clearly demarcated phases. Regarding saccades, this view is corroborated by the present findings, with the addition that initial saccades are somewhat shorter before their amplitude reaches its peak after the first second. There are two possible reasons why this initial rise of saccade amplitudes was not found by Antes. First, he recorded eye movements with eight frames per second only, thus it might be that most short saccades were not detected. Second, Antes calculated the values for temporal windows of two seconds, which might be too large an interval to reveal this short-lived effect.

From a theoretical viewpoint, the finding that initial saccades are shorter makes sense. When a novel image is presented to a participant, it is certainly efficient to first visually examine image elements in the near vicinity of the starting location before making saccades towards objects further off. Due to the resolution profile of the retina, promising fixation targets are easier to discern near the starting point than in the far periphery.

4 Experiment 2: Viewing task and stimulus repetition

4.1 Theoretical part

After having investigated the influence of stimulus size on various oculomotor measures in Experiment 1, two additional basic parameters will be studied in the present experiment. In the process of setting up a new scene perception experiment, the next questions after deciding on stimulus size might be, which task should be assigned to the participants, and whether it is admissible to repeat the stimuli. Viewing task, and thus the strategies that might be induced in the participants, are likely to influence various oculomotor measures. The impact of repeated presentation is yet another area worthy of further investigation, if only because researchers frequently include this in order to reduce the number of participants required to show an effect. Both of these parameters are of central importance for scene perception theory as well as experiment design requirements.

4.1.1 Viewing task

As introduced in chapter 2.1.4, "viewing task" can be understood in two ways. Either as completely different "things to do", in a sense that *stimuli* as well as *instructions* differ (e.g. reading vs. watching TV), or looking at the same stimulus with different viewing instructions (e.g. extracting the gist vs. memorising). For the present experiment, the second meaning of "viewing task" is relevant; it is defined according to the instructions given to participants prior to image presentation, but with the same stimuli. Since the time of the qualitative reports by Buswell (1935) and Yarbus (1967), it has generally been accepted that the task assigned for image viewing – and therefore the plan or strategy one adopts – is able to change spatial fixation distribution. For instance, participants viewing "An Unexpected Visitor" by the Russian painter I.E. Repin were found to fixate on different parts of the picture depending on the instructions given to them (Yarbus, 1967), with instructions such as free examination, estimation of the material circumstances or ages of the people, remembering the clothes or position of the people and objects in the room, and the like. However, I am not aware of any study with complex scene images which systematically varied the viewing task while holding the nature of the stimulus images constant in order to supplement these observations with quantitative data.

— Experiment 2: Viewing task and stimulus repetition

While there are no recent publications discussing possible effects of viewing task on the spatial distribution of fixations, there are at least some hypotheses regarding basic oculomotor measures. In a review article, Rayner and Pollatsek (1992) suggested that if a scene is shown to participants who are required to search for a target object, they will most likely show longer fixations and shorter saccade amplitudes, compared to other participants who are asked to look at the scene, but in anticipation of a recognition test. In a similar fashion, participants will exhibit slightly longer fixations and shorter saccades if they are asked to memorize all the objects they can see, compared to simply looking at the scene in order to extract the gist of it. This was attributed to the fact that, in the case of the difference in fixation duration, additional time is needed for memorization (Rayner & Pollatsek, 1992). In the present experiment, three tasks with different memory demands were contrasted: A free viewing condition (no explicit memory demands), a recognition task (intermediate memory demands) and a detail memorisation task (highest memory demands). As suggested by Rayner and Pollatsek (1992), I expected a similar pattern for the basic oculomotor measures, i.e. longer fixations and shorter saccades with increasing memory load. Based on pilot experiments in our lab (unpublished data), for the topology of fixations, I hypothesised that behaviour during free viewing would vary more between participants because a free viewing instruction gives them more option to choose different strategies, compared to the other, more specific tasks. Moreover, exploratory analyses of re-fixations as well as scanpaths will be reported, to investigate whether these measures are dependent on viewing task.

4.1.2 Repetition

It often seems desirable to repeatedly present images under different conditions. It will be analysed how stimulus repetition might influence – and thus confound – results of image viewing experiments. Concerning the effect of stimulus repetition on basic oculomotor measures such as fixation duration and saccade amplitude, the following papers are of central relevance. An early attempt into investigate the influence of stimulus repetition on eye movements with natural scene images was undertaken by Furst (1971). In his experiment, he used six different photographs (an aerial city view, two men, an old automobile, a masked face, an iceberg and a clockmaker). The images were presented five times for five seconds each in pseudo-randomised order. On average, the time between two viewings of the same image was approx. 2½ minutes. Furst found a steady decrease of fixation rate, which is tantamount to an increase of fixation duration, over the five trials. He interpreted this result as a sign of "habituation" of fixation rate. He also reported indications of another typical feature of habituation, namely recovery, i.e. fixation rate being slightly higher at the beginning of a trial compared to the end of the previous trial with the same image. In a study comparing amnesia patients with controls, Ryan, Althoff, Whitlow, and Cohen (2000) found a similar repetition effect, characterised by a decrease in sampling (i.e. lower fixation rate or longer fixation durations) of previously viewed scenes compared with new scenes. This was also the case in patients, thus it cannot be attributed to explicit memory for those scenes. Similarly, Flagg (1978) found an increase of fixation duration and a decrease of interfixation distance (i.e. saccade amplitude) in children repeatedly watching short TV scenes.

As introduced in Chapter 2.4.1, there are several studies suggesting that, during repeated viewing of complex images, some brief sequences of fixations and saccades tend to occur repeatedly (Groner et al., 1984; Noton & Stark, 1971a, 1971b; Pieters et al., 1999). However, these findings are controversial. Mannan et al. (1997b) found no stringent evidence of repeated scan patterns during viewing of scene images. However, in their study, similarity of fixated spatial locations was found to be high when sequence information was disregarded. Therefore, for the present study, I chose to concentrate on spatial fixation patterns irrespective of their temporal sequence, i.e. fixation topology. The specific hypotheses were: (1) Spatial fixations patterns of a participant, compared between the first and second viewing of a given image, are similar in tasks involving memory. Under these conditions, it was expected that viewers would tend to re-examine the same locations again during the second presentation in order to rehearse and verify the details memorized during the first viewing. In contrast, a lower similarity was expected in free viewing because participants might preferentially look at previously unexplored image areas. (2) The fixation patterns between different participants are more dissimilar when they are allowed to freely explore the images compared to more specific tasks like the recognition and memory task. Moreover, in an explorative fashion, it will be analysed whether stimulus repetition influences the number of re-fixations.

4.2 Methods

4.2.1 Participants

Fifty-four participants aged between 20 and 40 years ($M = 26.7$, $SD = 4.00$, 29 women and 25 men) participated and were assigned to three groups of equal size (henceforth referred to as group A, B and C). There was no statistically relevant difference either in age (Kruskal-Wallis test: $p = .877$) nor in male-female ratio (binomial test: $p = .93$) between the groups. All participants had normal or corrected-to-normal visual acuity and no strabismus, and were naïve as to the experimental hypotheses. Visual acuity was verified in a pragmatic manner after fitting the eye tracking equipment, by presenting a display with text in different sizes, and verifying that the participants were able to read the smallest line of print. The experiment was undertaken with the understanding and written consent of each participant. This work adheres to the Code of Ethics of the World Medical Association (Declaration of Helsinki), as well as to the ethical guidelines defined by the Swiss Academy of Medical Sciences SAMS („Integrity in Science: Guidelines of the SAMS for scientific integrity in medical and biomedical research and for the procedure to be followed in cases of misconduct", 2002). The study was formally approved by the local ethics committee (Ethisches Komitee der Universität Bern).

4.2.2 Stimuli

The stimulus images were colour photographs of real-life scenes. As scenes, I understand "specific views of the environment within which we are embedded" as recently suggested by Henderson and Ferreira (2004), therefore only images were selected that had been photographed from everyday vantage points and with approximately normal focal length, i.e. no aerial views, wide angle images or otherwise distorted perspectives. Forty-eight full colour images were selected, comprising landscape scenes, images of buildings, and populated city scenes (see Fig. 18 and Appendix C).

Fig. 18: Stimulus examples. The complete set of stimulus images is available in Appendix C.

4.2.3 Apparatus

The images were presented in a dimly lit room on a 19" cathode ray tube display with a resolution of 800 × 600, 24 bit colour depth, and a refresh rate of 85 Hz. Active screen size was 36 × 27 cm and viewing distance 70 cm, resulting in a viewing angle of 29 × 22°.
The eye tracking equipment was the same as in Experiment 1 (cf. Chapter 3.2.3). In addition to the described calibration procedures, this calibration was intermittently re-aligned, based on the fixations made in response to the central fixation point shown before each image. This procedure results in a gaze position accuracy of typically better than 0.5° over a trial block of up to 5 minutes.

4.2.4 Procedure

The experimental procedure is outlined in Fig. 19. The images were presented to the participants twice in randomised sequence, the second run after a delay of approx. 20 minutes. During this time, they were engaged in a different experiment (face perception). Viewing time was 5.5 s per image. Every image was preceded by a central fixation point (a dark dot of approx. 0.5° diameter on a 50% grey background) presented for 1.5 s, enforcing the same starting point for all participants and allowing for periodical recalibration of the eye tracker.
The three participant groups were assigned different viewing tasks. Group A was the free viewing group and received the instruction "Just look at the images". For them, the experiment was finished after the second presentation. Group B viewed the images in

— Experiment 2: Viewing task and stimulus repetition

preparation for a recognition test, which was administered after the second viewing. In this test, participants had to decide which of two images (one new, one shown before) they had seen before. Group C was instructed to expect a detail memory test, i.e. to detect a small image manipulation such as object deletion or colour change of any object. Recognition and memory task performance was not evaluated.

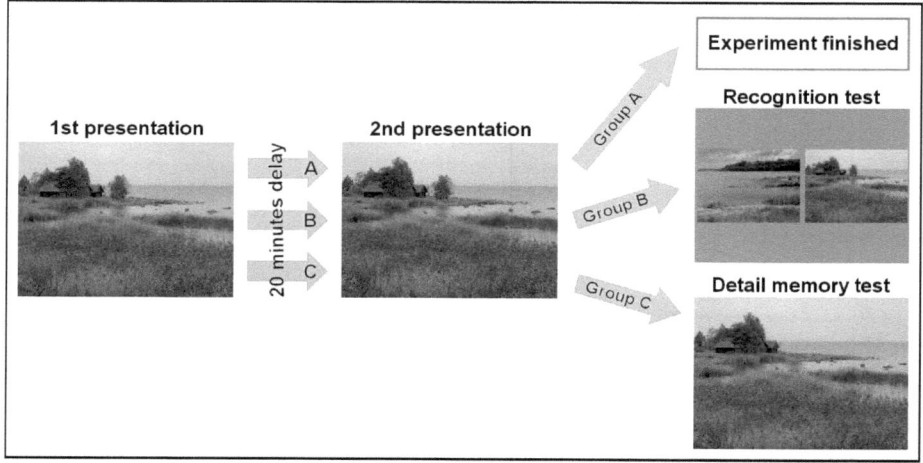

Fig. 19: Outline of the experimental procedure.

Each participant served for an experimental session of one hour at most, including instruction and debriefing; net time in the eye tracker did not exceed half an hour. In the instruction phase, the participants were given their specific task instructions. Then, they were shown an example image set, and – for Groups B and C – a sample recognition or detail memory test display, respectively.

4.2.5 Data analysis

For all analyses, only fixations with a duration over 100 ms and saccades with amplitudes over 0.75° were included. The fixation filtering procedure was necessary to eliminate fixations which are to be considered "non-cognitive" fixations. That is, such short fixation durations are not considered "regular fixations", but mostly occur before corrective saccades

(Carpenter, 1988). The saccade filter excluded short saccades in the range which is particularly difficult to parse for eye tracking systems, as concluded in Experiment 1.

Significance testing was performed using SPSS 12.0.1 (SPSS Inc., Chicago IL). In those cases general linear model analyses (GLM, McCullagh & Nedler, 1989) were used, deviations from sphericity were tested by Mauchly's Test of Sphericity and degrees of freedom corrected by a correction factor (method of Huynh-Feldt) if this assumption was violated.

4.2.5.1 Basic oculomotor measures

First, two basic oculomotor measures were analyzed, fixation duration and saccade amplitude. For every participant, mean fixation duration and mean saccade amplitude were computed for the whole viewing time and over all images, separately for the first and second viewing. Additionally, to illustrate the temporal course of these measures, mean fixation duration and mean saccade amplitude were calculated separately for ten time windows of 0.5 s duration, starting at the outset of the first fixation or saccade, respectively. Although the raw distributions of fixation duration and saccade amplitude are not normally distributed, it is safe to assume that the calculated saccade *means* and *medians* are. This normality assumption was verified for all analysed variables, by performing statistical normality tests on the variables themselves, as well as on their standardised residuals after the GLM statistics. Neither the Kolmogorov-Smirnov nor the Shapiro-Wilk test of normality found any significant deviation from normality.

4.2.5.2 Re-fixations

It might be that both factors, viewing task as well as stimulus repetition, cause a different amount of re-fixations. Thus, a measure was devised that expresses the incidence of repeatedly fixating a certain location with one or more intervening fixations at locations further away. In other words, a fixation was considered a re-fixation if it was located not more than 1.5° away from any earlier fixation *except the immediately preceding one*. The number of re-fixations was expressed in per cent of total fixations on the current image.

4.2.5.3 Scanpaths

For detecting scanpaths, i.e. repetitive sub-sequences of fixations identically made during the first and second viewing of a scene image, a new algorithm had to be implemented

— Experiment 2: Viewing task and stimulus repetition

(Matlab 6.1 Release 12.1, The MathWorks, Inc.). (1) In the first step, each set of fixations made by a single participant on a single image was searched for consecutive fixations that were less than 1.5° apart. With the rationale that such sequences are rather to be seen as "within-object re-fixations", such fixation pairs (or even triplets) were taken together as a single "object fixation", and their co-ordinates averaged. (2) In the second step, the two sets of fixations – or "object fixations" – made by a single participant on a single image were now searched for fixation sequences that were characterised by identical locations in the same temporal sequence. Two locations were considered "identical" if the distance between them did not exceed 1.5°. The resulting scanpaths were counted in categories of scanpaths with 2, 3, 4, 5, or 6 consecutive fixations. An example scanpath is illustrated in Fig. 20.

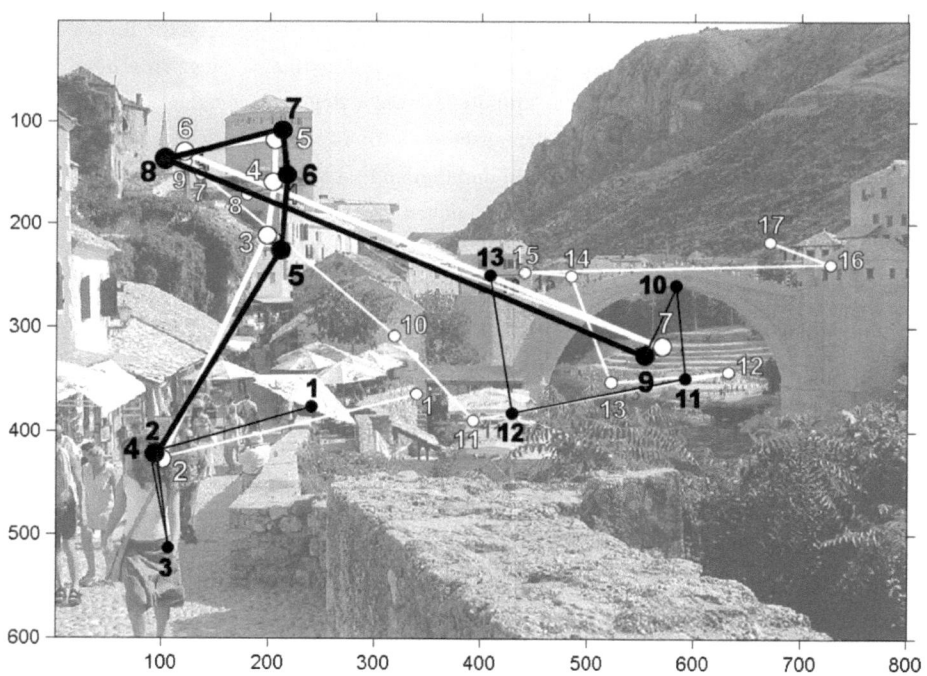

Fig. 20: Illustration of scanpath detection algorithm. Fixations and saccades of the two viewings of the same participant are shown in black and white. A six-fixation scanpath is highlighted by wider lines.

4.2.5.4 Similarity index

Furthermore, I was interested in how fixation topology was influenced by the two factors under scrutiny. To this end, the index of similarity I_s devised by Mannan et al. (1995) was adapted. The basic algorithm specifies the spatial congruence between two sets of locations $P_i(x, y)$ and $P_j(x, y)$ in a two-dimensional image, irrespective of the temporal sequence. It is defined as:

$$I_s = (1 - \frac{D}{D_r})100$$

where

$$D^2 = \frac{n_1 \sum_{j=1}^{n_2} d_{2j}^2 + n_2 \sum_{i=1}^{n_1} d_{1i}^2}{2n_1 n_2 (a^2 + b^2)}$$

n_1 and n_2 are the number of locations in the two sets, d represents the distances between a fixation in one set and its nearest neighbour in the other set, and a and b are the image dimensions. D is the calculation for the two empirical fixation distributions under scrutiny, whereas D_r stands for the value for two random sets with the same number of locations as in the empirical data, serving as the baseline. The similarity index is illustrated in Fig. 21.

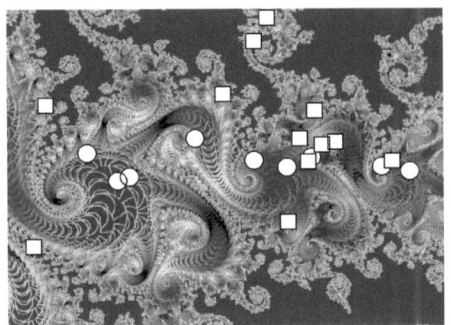
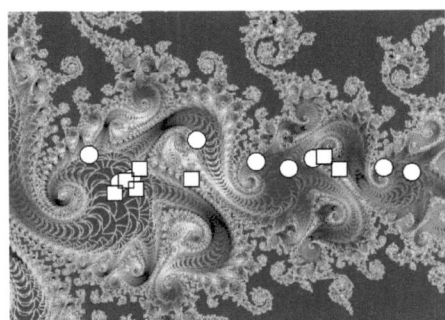

Similarity index $I_s = 10.19$ Similarity index $I_s = 60.12$

Fig. 21: Examples of fixation distributions resulting in low and high similarity index values. The circles and squares represent the locations fixated by two different participants.

More specifically, two different measures were defined. First, I wanted to know how similarly different participants look at the same images, and whether this inter-participant variation of spatial fixation patterns was influenced by viewing task and stimulus repetition. This measure will be termed inter-observer similarity index I_{sio}. To this end, a similarity index value was computed separately for each data cell (viewing task × image repetition). It was calculated for the fixated locations by any two participants in response to the same image; these values were then averaged over all possible pair comparisons.

The second measure, I_{srep}, was designed to specify whether the viewing task has any effect on the viewing pattern similarity between the first and second viewing, i.e. an intra-participant measure. It was obtained by calculating for each participant the similarity index for the two fixation patterns made on a certain image and averaging over all images.

Furthermore, baseline values expressing the overall similarity of traces I_{sb} for both measures were calculated. Basically, this is the same computation as for the other two similarity indices, but not comparing viewings of the same image, but of *different* images (see Mannan et al., 1995, for further details). This baseline is necessary to demonstrate that I_{sio} and I_{srep} express a significant similarity of viewing patterns which goes beyond a general similarity based on effects such as a commonly observed central tendency or similar image layouts.

In order to see whether the similarity between participants changes over time, both similarity indices I_{sio} and I_{srep} were also calculated in time windows of 1.5 s duration, a value suggested by Mannan et al. (1995). Contrary to them, however, I chose to evaluate non-overlapping phases.

4.2.5.5 Data visualisation

For data visualisation purposes (Fig. 30), the same algorithm as in Experiment 1 was used to create fixation density plots (see Chapter 3.2.5).

4.3 Results

4.3.1 Fixation duration

The values for mean fixation duration over the complete viewing time are given in Table 5. A repeated measures GLM analysis with the factors viewing task (between-group factor) and stimulus repetition (repeated factor) revealed main effects of viewing task ($F(1, 51) = 31.24, p < .001$) as well as stimulus repetition ($F(2, 51) = 3.645, p < .05$), but no interaction ($F(2, 51) = 0.995, p = .377$). Post-hoc tests (multiple pairwise comparisons) indicated that participants who had been assigned the recognition task fixated about 35 ms longer than the memory task group ($p < .05$), and 26 ms longer than the free viewing group, a difference which was statistically insignificant ($p = .151$).

In other words, a recognition task induces a viewing behaviour which is characterised by longer fixations than a detail memory task, with fixation durations under free viewing conditions lying somewhere in-between (Fig. 22). Moreover, when an image is shown a second time after several minutes of delay, fixations become considerably longer (about 16 ms), irrespective of the viewing task.

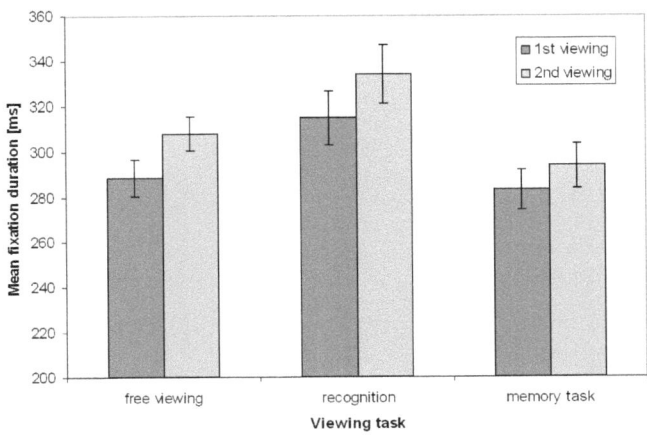

Fig. 22: Mean fixation duration over the whole viewing time. Error bars indicate the standard error of the mean.

— Experiment 2: Viewing task and stimulus repetition

Table 5: Fixation duration results. Values in brackets indicate the standard error of the mean.

	mean fixation duration [ms]	
	1st viewing	2nd viewing
free viewing	288.4 (8.063)	308.0 (7.615)
recognition	314.6 (11.77)	333.8 (12.95)
memory task	283.1 (8.762)	293.6 (9.933)

In Fig. 23, the temporal course of the change of fixation duration is shown (for the numerical data see Appendix D, Table D-1). On the one hand, this plot illustrates that fixation durations are steadily increasing with progressing viewing time. On the other hand, there are no clear signs of a "recovery" of fixation duration taking place during the 20 minutes between the two trials. In other words, the curves roughly continue as if there had not been any delay between the two presentations.

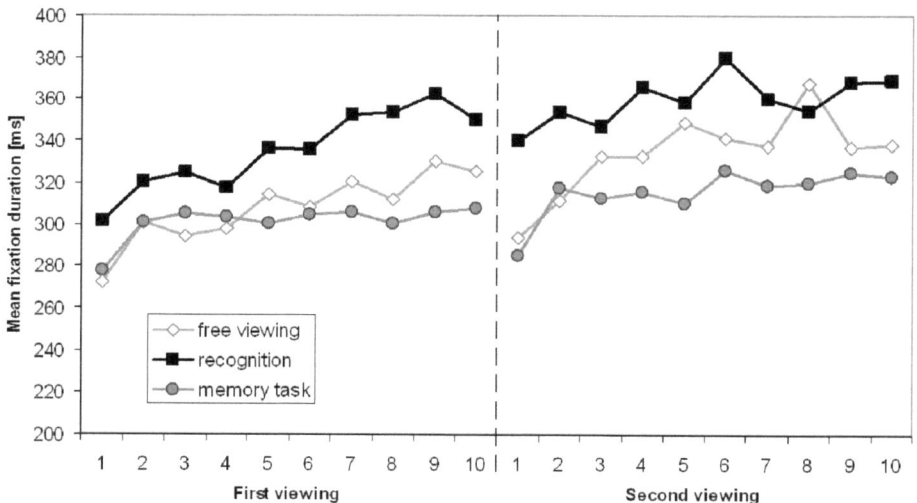

Fig. 23: Temporal course of fixation duration, with values calculated in temporal windows of 0.5 s. The dashed line represents the 20 minutes' delay between the two presentations.

To rule out the possibility that the cause for longer fixations during the second viewing is just an effect of fatigue, I analysed how fixation durations developed over the time of viewing one complete series of all 48 images, separately for the first and second viewing. The rationale was that if there is relevant fatigue involved, fixation durations should also become longer over the time of viewing all images, which was approx. ten minutes including two intervening re-calibrations and short rest phases. This, however, was not the case. The correlation of sequential image number (and thus time) with mean fixation duration was $r = -.008$ ($p = .958$) for the first viewing (i.e. no effect), and $r = -.287$ ($p = .049$) for the second viewing, that is a significant *decrease* of fixation durations. These values clearly argue against fatigue being the cause for the longer fixations during the second viewing.

4.3.2 Saccade amplitude

Mean values of saccade amplitude are plotted in Fig. 24, and listed in Table 6. In the GLM analysis, a main effect of repetition was found (repeated, $F(2, 51) = 77.79$, $p < .001$), but no effect of viewing task (between-group, $F(2, 51) = 0.106$, $p = .955$) and no interaction ($F(2, 51) = 2.902$, $p = .064$). That is, the second time a given image is explored under any of the three viewing tasks, the saccades are slightly but significantly shorter (about $0.4°$), whereas viewing task did not influence saccade amplitudes.

Table 6: Saccade amplitudes results. Values in brackets indicate the standard error of the mean.

	mean saccade amplitude [º]	
	1st viewing	2nd viewing
free viewing	5.359 (0.134)	4.825 (0.137)
recognition	5.318 (0.173)	4.948 (0.175)
memory task	5.207 (0.153)	4.933 (0.146)

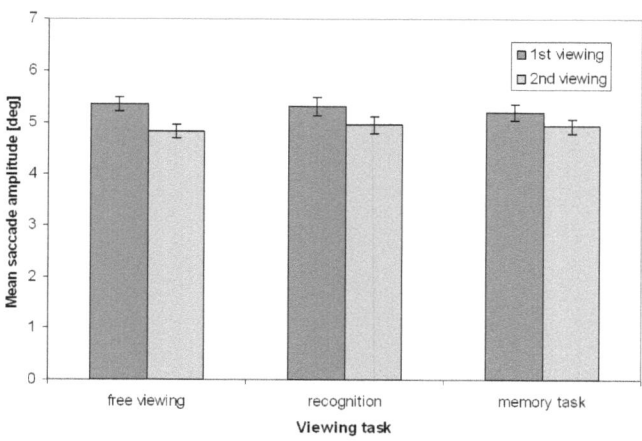

Fig. 24: Mean saccade amplitude over the whole viewing time. Error bars indicate the standard error of the mean.

The temporal course of saccade amplitude is displayed in Fig. 25 (for the numerical data see Appendix D, Table D-2). There is a prominent increase of saccade amplitude from the first to the second time interval, followed by the peak and a steady decrease afterwards. Thinking away the initial rise, the impression is similar to what has been found for fixation durations. The saccade amplitude change smoothly continues during the second viewing as if there had not been any delay in-between, that is no signs of any "recovery" of amplitudes after the delay.

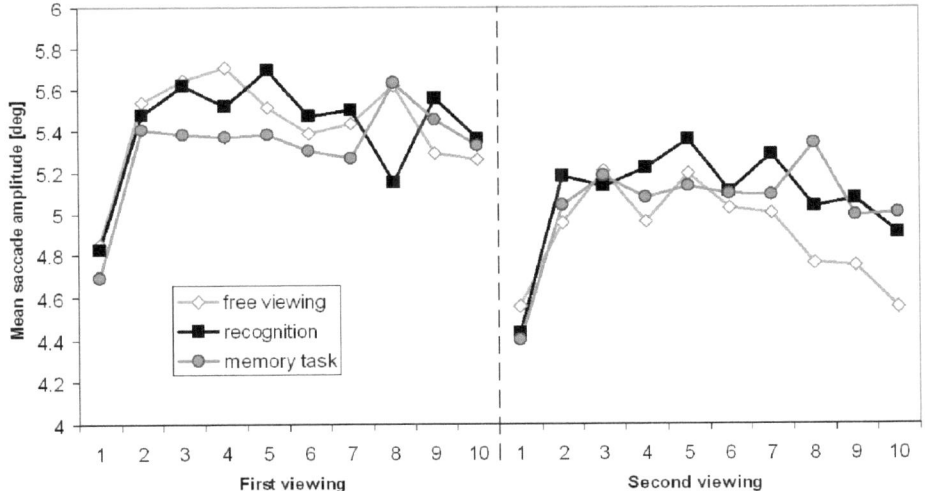

Fig. 25: Temporal course of saccade amplitude, with values calculated in temporal windows of 0.5 s. The dashed line represents the 20 minutes' delay between the two presentations.

— Experiment 2: Viewing task and stimulus repetition

4.3.3 Re-fixations

The percentage of re-fixations is given in Table 7 and Fig. 26. No effect of viewing task was found ($F(2, 51) = .209$, $p = .812$), but a significant effect of repetition ($F(1, 51) = 18.57$, $p < .001$). The analysis yielded no significant interaction between viewing task and repetition ($F(2,51) = .289$, $p = .75$). In other words, the second time a given image is viewed, slightly more re-fixations are made. Viewing task, however, was not able to change oculomotor behaviour in a way that different amounts of re-fixations had been the consequence.

Table 7: Re-fixations results. Values in brackets indicate the standard error of the mean.

	% of re-fixations	
	1st viewing	2nd viewing
free viewing	13.70 (0.526)	12.79 (0.425)
recognition	14.30 (0.581)	12.90 (0.603)
memory task	14.15 (0.421)	13.06 (0.457)

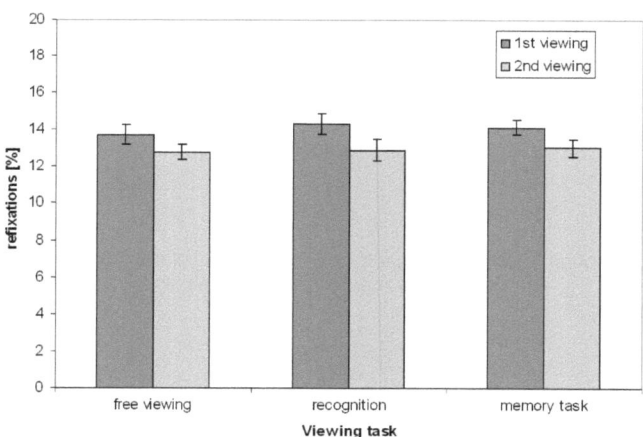

Fig. 26: Results of re-fixations analysis. Error bars indicate the standard error of the mean.

4.3.4 Scanpaths

The first question regarding scanpaths was whether a significant amount of scanpaths can be found at all during viewing of natural scenes. From the raw data (Appendix D, Table D-3) it can be seen that most of the scanpaths were very short ones; 84.8% were two-fixation sequences, and 12.7 % three-fixation paths. In Table 8, the absolute number of scanpaths per image, which was calculated within participant and within image, is contrasted with two control conditions. The "different participants" value expresses the number of scanpaths found when the calculations were performed within image, but comparing the first viewing made by a certain participant with the second viewing by a *different* participant. Accordingly, the "different images" value was obtained by counting scanpaths within participant, but comparing the first viewing of a certain image with the second viewing of a *different* image. Both calculations were iterated for all possible participant or image pairs, respectively. To determine whether a significant number of scanpaths was made, these three values (absolute number, control "different participant", and control "different image") were averaged over all three viewing tasks (values see Table 8, row "total", and Fig. 27). Two-tailed paired-samples t-tests indicated that the absolute number of scanpaths is significantly higher than both control calculations (absolute number vs. control "different participant": $t(53) = 10.88$, $p < .001$; absolute number vs. control "different image": $t(53) = 21.381$, $p < .001$). In other words, a small number of repetitive scanpaths was found, and this number is significantly higher than both control conditions.

Table 8: Scanpath analysis results. Values in brackets indicate the standard error of the mean.

	absolute number of scanpaths	control conditions	
		different participants	different images
free viewing	0.858 (0.048)	0.573 (0.019)	0.110 (0.008)
recognition	1.002 (0.076)	0.587 (0.023)	0.102 (0.009)
memory task	1.098 (0.087)	0.692 (0.020)	0.119 (0.008)
total	0.986 (0.043)	0.617 (0.014)	0.111 (0.005)

— Experiment 2: Viewing task and stimulus repetition

The second question was whether this small but significant number of scanpaths per image was influenced by the different viewing tasks. When looking at the values, there seems to be a tendency for more scanpaths being made during the two tasks involving memory (i.e. recognition and detail memory task). However, a one-way ANOVA indicated that this difference is not statistically significant ($F(2, 51) = 2.811$, $p = .069$).

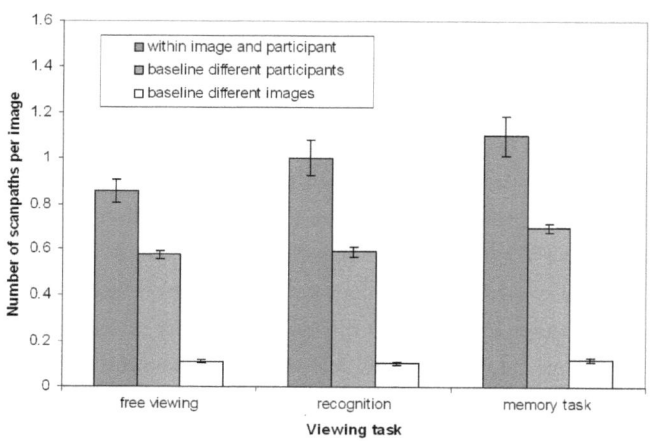

Fig. 27: Results of scanpath analysis. Error bars indicate the standard error of the mean.

4.3.5 Similarity index

As outlined in the methods section, two different similarity indices were computed: (1) An index I_{sio} termed „inter-observer similarity index", expressing the similarity of spatial fixation patterns between different participants, calculated separately for the three groups and for first vs. second presentation. (2) An "intra-participant similarity index" I_{srep} specifying the effect of the task on the viewing pattern similarity between the first and second viewing.

The I_{sio} values are plotted in Fig. 28 (left panel) and listed in Table 9. Two main effects were found. First, the repeated GLM task × repetition analysis revealed a main effect of viewing task ($F(2, 282) = 6.321$, $p < .01$), showing that the similarity in the recognition condition was higher than in the free viewing condition, with values in the memory task condition lying in-between (post-hoc tests: free viewing vs. recognition $p < .01$, recognition vs. memory task $p = .064$). Second, similarity was lower during the second viewing ($F(1, 282) = 7.199$, $p < .01$). There was no significant task × repetition interaction ($F(2, 282) = 0.952$, $p = .387$). That is, participants within the recognition group looked at the images in a more similar manner compared to the two other tasks; and in all groups, the similarity of fixation patterns is lower during the second viewing.

Table 9: Inter-observer similarity index results I_{sio} and baseline values I_{sb}. Values in brackets indicate the standard error of the mean.

	mean similarity I_{sio} [%]		baseline I_{sb} [%]	
	1st viewing	2nd viewing	1st viewing	2nd viewing
free viewing	25.29 (0.134)	19.49 (0.137)	-1.492	-4.197
recognition	28.67 (0.173)	26.98 (0.175)	0.8092	-0.1123
memory task	25.68 (0.153)	22.98 (0.146)	-2.655	-5.870

For I_{srep}, a one-way ANOVA statistic indicated a significant effect of viewing task ($F(2, 51) = 6.983$, $p < .01$), with the highest similarity in the recognition task (post-hoc tests: free viewing vs. recognition $p < .01$, recognition vs. memory task $p = .059$). The results are shown in Fig. 28 (right panel) and Table 10.

Table 10: Intra-observer similarity index results I_{srep} and baseline values I_{sb}. Values in brackets indicate the standard error of the mean.

	mean similarity I_{srep}	baseline I_{sb}
free viewing	29.14 (1.368)	-1.146
recognition	35.92 (1.679)	2.145
memory task	33.44 (1.200)	-1.480

— Experiment 2: Viewing task and stimulus repetition

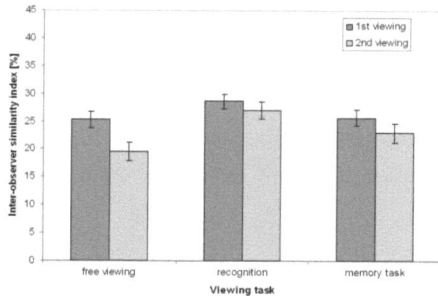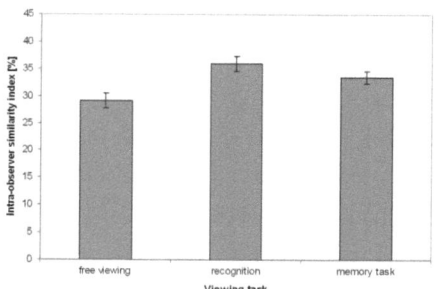

Fig. 28: Left panel: Inter-observer similarity index I_{sio}. Right panel: Intra-participant similarity index I_{srep}, i.e. first vs. second viewing. Error bars indicate the standard error of the mean.

As described above, baseline values I_{sb} for both measures, I_{sio} and I_{srep}, were computed. All values lay in the range of -5.8 and 0.8, and were significantly lower than the corresponding I_{sio} and I_{srep} values (two-tailed paired samples t-tests, $p < .001$ for all comparisons, for the values see Tables 9 and 10), underlining that these similarity indices indeed express similarities within an image that go beyond a general similarity based on effects such as a commonly observed central tendency or similar image layouts.

The values for the inter-observer similarity I_{sio} over time are plotted in Fig. 29 (top panel), separately for all factor level combinations. There was a main effect of time ($F(1.992, 561.6) = 484.2, p < .001$), with both changes from Phase 1 to Phase 2 as well as Phase 2 to 3 being significant ($p < .001$). There were no interactions with either viewing task or stimulus repetition (task: $F(3.983, 561.6) = .421$; repetition: $F(1.992, 561.6) = .217$).

For the intra-participant similarity index I_{srep}, the same pattern was observed (Fig. 29, bottom panel), i.e. a significant effect of time ($F(2, 102) = 275.2, p < .001$), but no interaction with viewing task ($F(4, 102) = .727$). The differences from Phase 1 to Phase 2 as well as Phase 2 to 3 were significant ($p < .001$ for both pairwise comparisons).

Experiment 2: Viewing task and stimulus repetition —

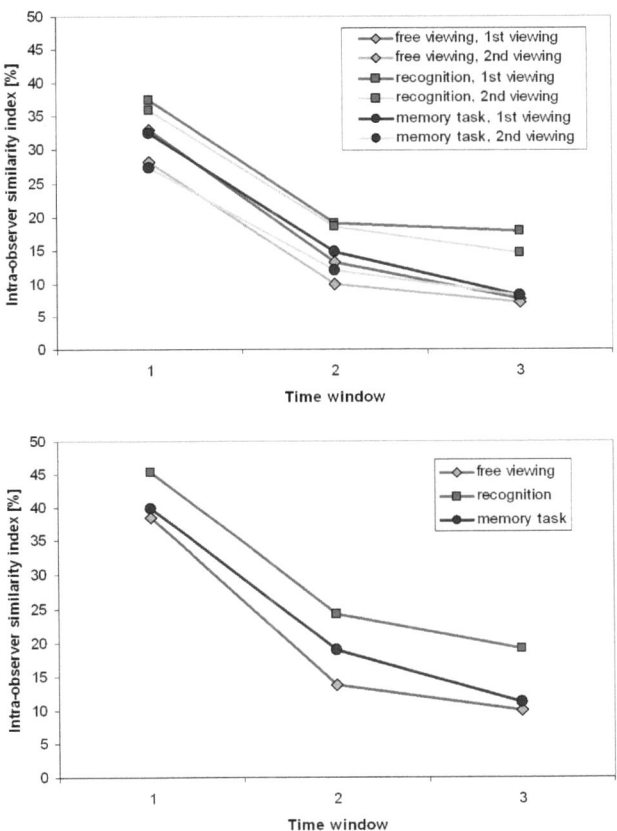

Fig. 29: Temporal course of similarity indices. Top panel: Inter-observer similarity index I_{sio}. Bottom panel: Intra-observer similarity index I_{srep}, i.e. first vs. second viewing.

To sum up, for both analysed measures I_{sio} and I_{srep}, the same temporal course was found, which is independent of the other factors, viewing task and stimulus repetition. All similarity index values were highest during the first 1.5 s viewing interval, with a steep decrease from the first to the second interval, and a further, less pronounced decrease from the second to the third interval.

4.4 Discussion

In the present study, I investigated the influence of the two factors viewing task and stimulus repetition on different aspects of oculomotor behaviour during scene perception. The results are discussed from two viewpoints. The first concerns the relation of the present results to earlier findings and theories, the second pertains to implications for further research in terms of experiment design.

4.4.1 Viewing task

With regard to oculomotor behaviour, the task assigned during a scene perception experiment is of central importance. Three different tasks were examined, free viewing, recognition, and a detail memory task. Of these three, the recognition task stood out.

First, it caused longer fixation durations than the two other tasks. It has been suggested that a task that causes a higher memory load (e.g. memorising vs. free viewing) leads to longer fixations (Rayner & Pollatsek, 1992). The present results suggest that this might not be the whole story. It was indeed the case that fixations in the recognition conditions were longer than in the free viewing condition. The suggestion that longer fixations would occur in a memorization task, compared to "looking for gist" (Rayner & Pollatsek, 1992), was supported by the present data. However, although the memory task used in this experiment is associated with higher memory demands than the recognition task, fixation durations were even shorter than in the free viewing condition. The present data suggest that the simple equation "higher memory load = longer fixations" is not always sufficient. Thus, it seems plausible to assume that other factors, such as the *viewing strategy* the participant adopts, have an impact on fixation duration. For instance, in a recognition task, it might be advantageous to select few image elements – preferably meaningful ones – that are characteristic for a given image, and fixate them for longer times in order to memorize them better. In a detail memory task, it might be better to find a trade-off between well memorising details and looking at as many objects as possible in the available time. For an illustration of this idea see Fig. 30.

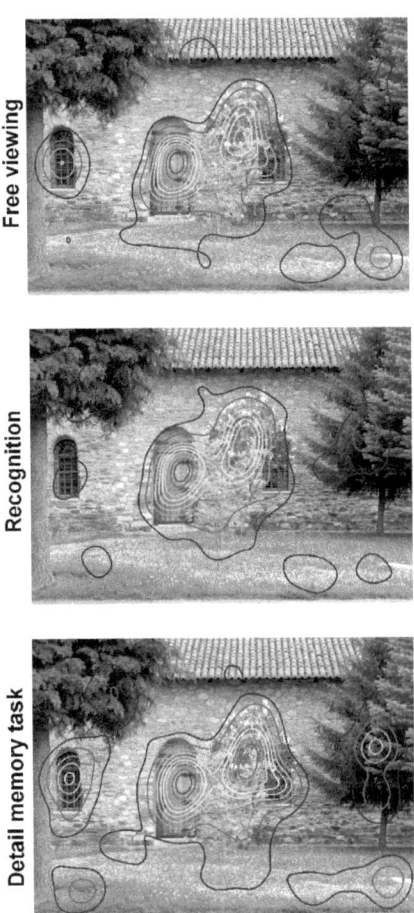

Fig. 30: Fixation density plots over all observers on an example image. In the recognition condition, a stronger tendency to concentrate on the two central details and thus make less fixations on more peripheral locations can be seen. Furthermore, the greatest number of distinct image regions were fixated in the detail memory task condition.

However, Rayner and Pollatsek (1992) also suggested that shorter saccades would occur in a memorization task compared to "looking for gist", a contention I cannot confirm with the

— Experiment 2: Viewing task and stimulus repetition

data from this experiment; the viewing task manipulation did not affect saccade amplitude in any significant way. Unfortunately, this aspect is difficult to discuss because, although spatial layout and display size are supposed to be crucial factors, they are not specified in sufficient detail in Rayner and Pollatsek's article (1992). In Experiment 1, I concluded that, apart from image size, saccade amplitudes are rather indifferent to other factors such as object spacing in the image, stimulus properties, or measurement equipment specifications. I can now supplement these findings with the observation that this is also true for viewing task.

Second, I found that the fixation pattern similarity between the first and second viewing was highest in the recognition task. This result can be interpreted in a similar manner, that is, based on viewing strategies. Under free viewing conditions, there is no reason to look at the same objects again during the second presentation of a certain image. For a recognition task, however, it might be adaptive during the second viewing to "visually rehearse" those characteristic items selected during the first run, thus leading to a high similarity between the two runs. As the probably most useful strategy for the memory task is to try and grasp as many objects as possible, the second presentation will be used to look for memorisable items not fixated before. Similar to free viewing, this leads to a lower similarity value.

Third, it was found that different participants looked at the images in a (spatially) more similar way when they had been assigned a recognition task, compared to the free viewing and detail memory task. It seems that the recognition task forces a similar strategy upon different participants, which also results in similar oculomotor behaviour. Additionally, the selection of those few "characteristic" objects necessary to solve the recognition task is supposedly guided by processes that are not strongly idiosyncratic. On the one hand, this might be the processes guiding visual attention based on bottom-up information. A prominent theory of pre-attentive, bottom-up visual attention, the *saliency map approach* (Itti & Koch, 2001; Koch & Ullman, 1985), assumes that selection of attentional targets is based on a small number of separate feature channels (e.g. luminance contrast, colour, and orientation), which are evaluated in a centre-surround manner. Afterwards, they are combined into a "saliency map", which is a topographical map containing the merged local maxima arising from the single feature channels (cf. Chapter 2.2.2). By definition, the momentary maximum on the saliency map is the most "conspicuous" location. On the other hand, in scene perception, "scene schemas" as suggested by different researchers (Antes et al., 1981; Biederman, 1972; Henderson, 1992a) might contribute to the unifying effect of the recognition task, insofar as they restrain the image locations from where the crucial objects are preferably selected. The *scene schema hypothesis* proposes that scene recognition depends on the early activation of a few typical scene representations in memory to guide information extraction (Schyns & Oliva, 1994).

A brief side note regarding similarity index values: When the absolute values from the present experiment are compared with the data reported by Mannan et al. (1995), it might seem surprising that their values of the inter-observer similarity index were considerably higher, with $I_s = 63$ for the initial 1.5 s and $I_s = 40$ for the full 3 s period. This difference is attributed to the fact that the images they used mostly depict either people ("Old woman in room", "Man on a horse", "Two dancers" etc.) or scenes with buildings, and sometimes human artefacts. In such images, there are fewer potential "centres of attention". According to common experience, human faces are extremely powerful attractors, resulting in exceptionally high similarity index values. Moreover, we have demonstrated earlier that, compared to colour images, achromatic images further reduce the number of spots capable of attracting visual attention, for there are no chromatically defined foveation targets, potentially further increasing the similarity values (Jost et al., 2005; von Wartburg et al., 2005).

With regard to experimental design, some recommendations arise from the results reported here. On the one hand, I conclude that it is generally preferable to assign participants a well-defined task because otherwise the free viewing instruction leaves them at liberty to choose any (explicit or implicit) strategy they like, which might lead to unpredictable behaviour and increases the variance of spatial fixation patterns between participants. On the other hand, when choosing a proper task to be assigned in a scene viewing experiment, it should be borne in mind that memory load is not the only important aspect. The strategy the participants adopt in order to appropriately solve the task are equally important.

4.4.2 Stimulus repetition and temporal course

In the current experiment, the repeated presentation of stimuli led to several effects. First, if a series of images are presented for the second time, longer fixation durations are observed (i.e. a lower fixation rate), irrespective of the viewing task. This finding is consistent with an early finding reported by Furst (1971). While repeatedly looking at photographic images, participants displayed a behaviour interpreted as "habituation" of fixation rate: The fixation rate steadily decreased over the five trials, which is tantamount to an increase of fixation duration. However, contrary to Furst (1971), I did not find conclusive evidence for any substantial "recovery" of fixation rate. When plotting the temporal course of fixation durations of the first and second viewing side-by-side (cf. Fig. 23), the graph rather looks comparable with what would be expected if participants had viewed an image continuously for eleven seconds. That there is a continuous increase of fixation duration over a longer time of viewing has been demonstrated by Antes (1974). Thus, a repeated presentation of an image results in a similar temporal profile of fixation durations as a longer presentation of the same image, at least for relatively short delays as applied in the present study. The same seems to be valid for the temporal course of saccade amplitudes (Fig. 25), just in the other direction, i.e. saccade amplitudes decrease over time, with no distinguishable "recovery" after a delay. A similar decrease of fixation rate has also been reported for repeatedly presented TV scenes (Flagg, 1978), suggesting that this effect is not restricted to still image viewing. Interestingly, the observed repetition effects do not seem to depend on explicit memory for scene contents, as suggested by Ryan et al. (2000). They found an analogue effect even in patients with severe amnesia who had no explicit (verbal) memory for the scenes. This suggests that this is either an effect of implicit memory, or of visuo-spatial memory, which might have been spared in the patients. Taken together, Ryan *et al*'s report and the findings from the current experiment suggest that the decrease of fixation rate over repeated presentations is not a question of base-level habituation, but rather a more enduring effect of (possibly implicit) memory.

An alternative explanation for the finding that fewer (but longer) fixations are made during the second presentation is fatigue. However, I deem this implausible for the current study, as the experiments are no more exhausting than normal viewing behaviour for instance while watching TV or working at the computer. This claim is supported by the finding that, over the time an participant viewed all 48 images, no increase of fixation durations was found, neither the first nor the second time the images were presented. Thus I feel safe in concluding that fatigue is irrelevant here.

The second main finding was that saccades during the second viewing were on average slightly shorter than while viewing a novel image. This suggests that Flagg's finding (1978) that saccade amplitudes continually decrease in children repeatedly watching short TV scenes can be generalised to static natural scene images. A tentative explanation for this decrease of saccade amplitudes could be that the spatial layout of the image, and thus the arrangement of "interesting" objects, is already known when the participants see a complex visual stimulus for the second time. Thus, the need for long saccades which are made in order to capture the overall image structure is reduced, and the important objects can be foveated in an optimised sequence.

In Experiment 1 it was found that image size was the dominant factor with regard to saccade amplitudes. The present data suggest that stimulus repetition is one of the few factors which also has a significant influence on mean saccade amplitudes. From this it can be deduced that total viewing time per image will have a similar effect. As saccades tend to become shorter over time, longer exposure durations will lead to lower mean (or median) saccade amplitude values.

Third, when comparing spatial fixation patterns, the between-participants similarity was lower during the second viewing. If it is correct that, as I suggested above, viewing an image twice is comparable to viewing the same image once for double the time, and that the first few seconds are predominantly governed by pre-attentive, bottom-up effects (Mannan et al., 1995), this would not be surprising. While these bottom-up effects are not strongly idiosyncratic, the top-down effects that exert their influence only later in a trial (i.e. strategies, memory) are highly so. Or, in other words, the longer participants are allowed to look at an image – with or without intervening break – the more dissimilar will their fixation patterns be, as behaviour becomes more strongly governed by idiosyncratic top-down processes.

Considering the time course of the inter-participant and the intra-participant similarity index, both measures yielded a clear effect: The similarity values markedly decreased from the first to the second phase and showed a further, but less steep, decrease towards the end of the trial. This both validates and extends the findings of Mannan et al. (1995) who displayed their images for 3 s only. The present data support the argument that during the first 1.5 s, attentional deployment is predominantly under bottom-up control and is thus less idiosyncratic. It can now be added to this the insight that this temporal progression is largely independent of the task assigned for image viewing, at least for the three tasks applied in the present study.

Thus, from a methodological standpoint, it became clear that it is in most cases not admissible to apply repeated presentations of the same stimuli. Unless, of course, stimulus repetition is itself at issue. Moreover, it seems advisable to use a viewing duration of more than three seconds when oculomotor aspects are under scrutiny, as several measures behave differently during the earliest part of viewing.

5 Summary and Conclusions

In this thesis, my scientific work on the influence of basic parameters on oculomotor behaviour during scene perception is reported. Eye movements, including complex measures based on the topology of fixations as well as more basic oculomotor measures, are very sensitive indicators of different visuo-cognitive states. Thus, a small change in experimental manipulations can evoke large effects, a fact which should be taken into account with regard to both oculomotor research methods and theoretical models of visual attention and oculomotor control.

5.1 Theoretical aspects

The results of the experiments conducted for this thesis allow me to further specify some of the processes taking place during the time a person looks at a scene image. In a typical scene image viewing experiment, when a new image appears on the display, the participant's gaze is already directed to a certain location on the display. Usually, the participant has been informed beforehand what kind of stimuli s/he is to expect, how much time there is available for inspecting the images, and what s/he has to do with the image information (i.e. the viewing task).

As outlined before (Chapter 2.4.3), the typical course of events commences with a first fixation, during which the spatial layout of conspicuous regions is acquired by bottom-up processes. According to the saliency map approach of visual attention (Itti & Koch, 2000, 2001; Koch & Ullman, 1985), this process is based on a multi-scale centre-surround evaluation of visual information in different feature channels, such as brightness, colour, or orientation. In this process, the visual system also registers the borders of the area of information relevant for the task to solve, or, in other words, the size of the presented image. This is very effective in avoiding "useless" fixations beyond the image borders; in Experiment 1, it was found that only an evanescent number of fixations were made outside the actual image area.

Based on the information collected during the first fixation, the target of the subsequent fixation is selected, and the first saccade is planned accordingly. Due to the resolution profile of the retina and associated low-level neural structures, potential targets in the vicinity of the starting point (i.e. the current fixation) are more likely to be correctly judged as "useful" targets. Therefore, it is an efficient strategy of the visual system to first make saccades towards promising locations near the starting point. In the present data, this is apparent in the temporal graph of saccade amplitudes, which shows that initial saccades tend to be shorter than later ones.

That oculomotor behaviour seems to be of a different quality during the first one or two seconds is a phenomenon which does not only pertain to saccades. During the initial viewing phase, fixation durations are also shorter. The data further suggest that the peak of saccade amplitudes is reached after approx. one to two seconds. This might be interpreted as part of a "survey scan phase" (Antes, 1974; Buswell, 1935; Karpov et al, 1968) during which – after having scanned the first few promising targets around the starting location with short fixations – more peripheral, salient spots are checked for useful content.

Moreover, measures characterising the similarity of spatial fixation pattern between and within participants also exhibit prominently higher values during the initial one or two seconds. This reflects the theory that, when we begin to explore a novel image, the processes initially guiding human gaze are governed by pre-attentive, bottom-up effects (Mannan et al., 1995). These processes are thought to be less idiosyncratic than the top-down effects that exert their influence only later in a trial, thus inter-individual differences in viewing behaviour are smaller.

After these first one or two seconds, various top-down effects begin to exert their influence on oculo-motor behaviour. The present work revealed that, after a peak has been reached, saccade amplitudes slowly and continually decrease, a tendency which even continues when the same image is repeatedly shown with intervening delays. This suggests that, based on the spatial layout of objects which becomes more and more detailed with each fixation, the visual system continually optimises the sequence of spatial fixation locations. Therefore, the main task shifts from gathering spatial information towards object recognition and – if the viewing task requires it – memorisation. Thus, in the same time, fixation durations continually increase towards the end of viewing time.

As far as top-down aspects are concerned, the results presented in this thesis mainly pertain to the viewing task, and thus to the strategies as they are instigated by different instructions given to participants. Contrary to earlier suggestions (Rayner & Pollatsek, 1992), effects of the viewing task on the duration of fixations cannot only be explained based on memory load. In Experiment 2, the longest fixations were not measured during the task with highest memory demands, but in the recognition condition. These results could be explained as follows. In a free viewing task, participants are not required to reason much about the image, thus overall fixations are short. In a recognition task, however, it seems to be useful to search the image for objects that are characteristic for a given image, and fixate them for longer times in order to memorize them better. In a detail memory task, then, it might be optimal to scan as many objects as possible in the available time. This strategy, however, necessarily reduces the time available for memorising the image details.

This difference in strategies is also reflected in the similarity of spatial fixation patterns between the first and second viewing. For a recognition task, it is adaptive during the second viewing to "visually rehearse" those image elements selected during the first run. Therefore, the viewing pattern similarity in the recognition condition is higher than in the other conditions.

A second, relevant top-down aspect is memory. At least for a certain time, the information gathered during scene perception remains stored in memory, or – to be more precise – to a large extent in *visuo-spatial* memory. Therefore, the effects of time observed during a single image trial, such as increasing fixation durations as well as decreasing saccade amplitudes and similarity values, also appear when comparing two presentations of the same scene with a considerable delay in-between.

5.2 Methodological aspects

From these conclusions regarding visuo-cognitive processes, several recommendations regarding methodological aspects can be derived. According to my personal experience since the time I begun research on scene perception, the central questions that arise early in the process of setting up a scene perception experiment concern image size, exposure duration, whether the scenes should be shown in colour or black-and-white, the viewing task that is assigned to the participants, and whether it is appropriate to repeat stimulus presentation in order to reduce the number of participants necessary to show an effect. These questions coincide with the topics tackled in earlier work of mine and the present thesis.

Image size: The data from Experiment 1 strongly suggest that, in scene viewing studies, it is important to display the stimuli in reasonable size, that is at least 20°. With smaller images, the ratio between measured amplitude values and measurement error becomes unfavourable, as mean saccade amplitudes scale with image size. Moreover, I have pointed out that it seems advisable to exclude saccades below approximately one degree of visual angle. With current eye tracking equipment, movements in this range are difficult to differentiate from micro-movements that are not to be considered "regular" saccades. Using larger images minimises possible distortion caused by this filter procedure. But there is also an upper limit to image size, at least when the participants' heads are fixed during the eye tracking session. Compared to real free viewing conditions in everyday situations, it seems unnatural to make such large saccades without moving head and/or body. I estimate the upper limit to be around 35° horizontally; in Fig. 9 it can be seen that on the largest images, less fixations are placed near the border of the image. This might indicate that this size, i.e. 34°, is already a little bit too large.

Exposure duration: If the researcher is interested in overall measures such as mean saccade amplitude or fixation duration, it is recommendable to apply a viewing duration of more than three seconds. It has to be borne in mind that several measures behave differently during the earliest part of viewing. With short presentation times, this initial phase would have an inappropriately large influence on overall measures.

Colour: This issue was not considered in the present work, but in earlier work (von Wartburg et al., 2005). It was concluded therein that it is better to present full-colour images instead of grey-scale ones. Applying achromatic images reduces the ecological validity of results, and leads to a smaller variation between spatial fixation behaviour between participants.

Viewing task: The present results suggest that the instructions given to the participants of a scene perception experiment can have a strong impact on several basic and derived eye movement measures, except saccade amplitude measures. It should be considered that the strategies that are evoked by a certain instruction might override or interact with effects of memory demands and the like. Furthermore, it is generally preferable to assign participants a well-defined task – such as memory tasks, ratings etc. – instead of just leaving them at liberty to choose any explicit or implicit strategy they like. A free viewing instruction increases the variance of spatial fixation patterns between participants and might lead to unpredictable behaviour as the researcher does not know what the participants actually do. In addition, the current results underline that comparisons of results between different studies might be inappropriate in case different explicit viewing instructions were involved. This is especially important when fixation durations and/or the spatial distribution of fixations are at issue. For studies investigating saccade amplitudes, however, viewing task seems irrelevant.

Stimulus repetition: From the present results, it becomes evident that stimulus repetition has the potential to confound results obtained in scene perception studies, even for mean saccade amplitudes, which have been shown to be rather robust against other experimental manipulations except image size. Thus, it is important to either completely avoid repetition, or at least to safeguard that conditions are appropriately balanced.

5.3 Outlook

In this thesis, several questions regarding eye movements during scene perception have been investigated, and important insights have been gained. I hope that my work will help to build better experiments in the future, and that my findings and conclusions initiate further research. It is the usual course of events in the sciences that every experiment, every finding, and every conclusion are only temporary, call for replication, and raise further questions. One of the more immediate questions is, for instance, what exactly leads to the change in several oculomotor measures during repeated presentations of the same stimulus, or – more generally – to the change over time. Is it explicit, declarative knowledge, or rather an implicit memory faculty? Is it possible to show, for example, that patients with impaired visuo-spatial memory after brain damage scan scene images differently than healthy controls? In a similar fashion, it would be interesting to investigate how brain damage to more frontal areas, associated with strategic cognitive processes, change fixation patterns during scene perception.

The present results make it more evident that many basic parameters in scene perception studies might be more important than previously thought, at least as far as scene perception research with images on computer displays is concerned. In extrapolation, this suggests that all these factors, and several more that are irrelevant when using 2D computer displays, will also have to be considered in the next, logical step, that is when taking scene perception studies from the laboratory into the real world.

Name index

—A—
Andrews, 39
Antes, 13, 18, 23, 33, 36, 39, 40, 41, 57, 58, 67, 92, 94, 97
—B—
Balint, 14
Biederman, 23, 34, 92
Brandt, 2
Brefczynski, 20
Buswell, 2, 17, 33, 35, 36, 40, 67, 69, 97
—C—
Carpenter, 13, 75
Cavegn, 11
Curcio, 6
—D—
Deubel, 27, 29
—E—
Enoch, 39, 61, 62, 63, 65
—F—
Findlay, 3, 29
Fischer, 9, 11, 12, 13
Flagg, 71, 94
Friedman, 33
Furst, 70, 93, 94
—G—
Gilchrist, 11
Gould, 35
Groner, 31, 71
—H—
Henderson, 3, 8, 16, 17, 21, 23, 28, 29, 32, 34, 35, 38, 39, 43, 59, 65, 72, 92
Hodgson, 28
Hoffman, 27, 28
—I—
Irwin, 15, 38
Ishihara, 43
Itti, 19, 23, 24, 32, 92, 96
—J—
Jost, 32, 93
—K—
Karpov, 17, 67
Koch, 24, 25, 32, 92, 96
—L—
LaBerge, 26
Land, 39
Legge, 38, 65
Lindman, 48
Liversedge, 5, 6
Loftus, 15, 17, 33, 38, 39, 59

—M—
Mackworth, 33
Maioli, 29
Mannan, 30, 31, 36, 37, 39, 59, 71, 77, 78, 93, 95, 97
Matin, 7
McCullagh, 47, 75
Morrison, 38, 65
Müri, 12, 25, 32
—N—
Nelson, 7
Niebur, 24
Noton, 8, 30, 31, 71
—O—
Ouerhani, 25, 32
—P—
Parkhurst, 25, 60
Pierrot-Deseilligny, 12
Pieters, 31, 71
Posner, 27
Pratt, 26
—R—
Rayner, 6, 7, 10, 13, 16, 17, 35, 36, 38, 39, 65, 70, 90, 91, 92, 97
Rizzolatti, 28
Ryan, 71, 94
—S—
Saida, 7, 60
Salthouse, 15, 16, 38
Sanders, 27
Saslow, 11
Schiller, 9
Schneider, 27
Schyns, 23, 34, 35, 93
Shioiri, 60
Starl, 14
Stevens, 21
—T—
Tatler, 39
Treisman, 21, 22, 24, 25
—V—
van Diepen, 16, 58
von Wartburg, 3, 17, 41, 42, 61, 67, 93
—W—
Wandell, 5, 6
Wright, 19, 20, 26, 27
—Y—
Yarbus, 2, 9, 30, 35, 69

Subject index

—A—
achromatic, 93
amnesia, 71, 94
analogue zoom lens model, 20
attention, 19
  control, 23
  covert, 27, 29
  overt, 27, 29
  premotor theory of, 28
  visual, 2, 8, 19, 20, 21, 27, 28, 29, 32, 96
attentional control
  bottom-up, 21
  top-down, 21
awareness
  visual, 20

—B—
Balint syndrome, 14
binding, 22, 25
bottleneck, 22
bottom-up, 17, 21, 24, 25, 31, 32, 35, 36, 92, 95, 96, 97

—C—
capacity limitation, 20, 22
centre-surround, 24
cognitive economy, 32
cognitive processing, 15
colour, 17, 93, 99
computational modelling, 19, 25
concept driven, 21, 23
cone density, 5
cones, 4
conjunctive search, 22, 23, 24
contrast, 36
control models, 10
  direct, 10
  global, 10
  indirect, 10
cortex
  visual, 24

—D—
data driven, 21, 22
decoupling hypothesis, 28
destination indexing model, 28
detail memory task, 92, 98
detailed looking phase, 17, 36, 40, 67
discriminability, 17, 21
drift, 9, 14

—E—
ecological validity, 2
edge density, 36
endogenous selection, 23
evolutionary significance, 19
exogenous selection, 22
experience, 25
experiment design, 67, 93, 95, 98
exposure duration, 99
eye movement, 1, 4, 7, 14, 19, 21
  conjugate, 9
  control models, 10
  parsing, 40, 66
eye tracking, 2, 40

—F—
familiarity, 21
fatigue, 94
feature integration theory, 21, 24, 25
feature map, 24, 25
fixation, 7, 10, 14, 32, 35, 96
  duration, 10, 15, 18, 70, 79, 93
  spatial pattern of, 33, 35, 95
  system, 14
  topology of, 30
fMRI, 20
fovea, 4, 27
foveation, 7, 9, 10, 14, 27, 36
foveola, 5
free viewing, 16, 92, 97
frontal lesions, 17

—G—
ganglion cells, 6
gap paradigm, 11, 14
goal-driven, 26
grey-scale, 17, 93, 99

—H—
habituation, 71, 93, 94

—I—
image
  perception, 13, 16
  size, 38, 65, 96, 98
  viewing, 38
indexing, 28
information density, 17
informativeness, 33, 36
inhibition, 25
inhibition of return, 25, 36
IOR, 25, 36

—K—
kurtosis, 49

—L—
lateral geniculate nucleus, 24
learning, 24
lgn, 24
luminance, 17, 37

—M—
memory, 70, 90, 95, 98
  explicit, 94
  implicit, 94
  short-term, 20
  visuo-spatial, 98
memory task, 16
meta analysis, 58
methodology, 67, 93, 95, 98
microtremor, 9, 14
minimum pause time, 15, 38
—N—
natural scene, 30, 34
nystagmus, 9
  optokinetic, 9
  vestibular, 9
—O—
object
  recognition, 7
  spacing, 65
ophthalmoplegia, 11
—P—
paradox of intelligent selection, 32
parafovea, 5
parafoveal deadzone, 13
perceptual span, 6
periphery, 5, 7, 32, 34, 68
pop-out, 22, 23, 24, 25
pre-attentive, 24, 25, 26, 92, 95, 97
premotor theory, 28
prioritised processing, 19
pursuit, 8
—R—
reading, 7, 13
receptors, 4, 5
recognition, 16, 31, 92, 98
recovery, 71, 94
re-fixation, 36, 71, 75, 84
regions of interest, 33
repetition, 69, 70, 93, 100
retina, 4, 24, 27, 68, 96
retinal periphery, 7
rods, 4
ROI, 33
—S—
saccade, 7, 10, 14, 27, 29, 36, 38
  amplitude, 10, 13, 38, 39, 65, 70, 82, 94
  amplitude distribution, 66
  anti-, 12
  anticipatory, 12
  corrective, 13, 74
  express, 12
  fast regular, 12
  inhibition, 14

  intentional, 12
  intentional visually guided, 12
  latency, 8, 11, 12, 13
  memory guided, 12
  micro-, 14
  predictive, 12
  reflexive, 12
  reflexive visually guided, 12
  secondary, 13, 74
  slow regular, 12
  spontaneous, 13
  temporal course of, 40
  voluntary, 12
saliency, 21, 24, 25, 33, 35
saliency map, 25, 26, 32, 92, 96
scaling, 65
scan pattern, 32
scanpath, 8, 30, 31, 71, 75, 85
  global, 31
  local, 31
scene
  perception, 13, 29, 32, 35, 39, 65
  schema, 25, 34, 36, 92
  schema hypothesis, 23, 35
schema, 21, 23
search, 16
semantic, 34
sequential attention model, 28
similarity index, 30, 77, 86, 93, 95, 97
skewness, 49
smooth pursuit, 8, 14
spatial
  frequency, 37
  resolution, 5, 6
spotlight model, 20, 26
  analogue, 20
  discrete, 20, 26
stimulus driven, 21, 26
strategy, 16, 21, 23, 25, 69, 90, 92, 95, 97
superior colliculus, 24
survey scan phase, 17, 36, 40, 67
—T—
temporal course, 40, 67, 93, 94, 95
top-down, 16, 21, 23, 24, 25, 31, 34, 36, 95, 97
training, 25
Troxler effect, 9
—U—
useful field of view, 6
—V—
V1, 6, 7, 24
V4, 20
V5, 24
vergence, 9, 14

Subject index

vestibulo-optic reflex, 8
viewing task, 13, 16, 23, 35, 39, 69, 90, 93, 97, 99
vision
  photopic, 4
  scotopic, 4
  stationary, 9
visual
  cortex, 20, 24
visual attention, 2, 8, 19, 20, 21, 27, 28, 29, 32, 96
  activity distribution model of, 26
  control mechanisms, 24
  selective, 19
  shifts of, 21
  two-component model of, 21, 25
visual system, 4
volition, 21, 23, 25
VOR, 8, 14

References

Andrews, T. J., & Coppola, D. M. (1999). Idiosyncratic characteristics of saccadic eye movements when viewing different visual environments. *Vision Research, 39*(17), 2947-2953.

Antes, J. R. (1974). The time course of picture viewing. *Journal of Experimental Psychology, 103*(1), 62-70.

Antes, J. R., Penland, J. G., & Metzger, R. L. (1981). Processing global information in briefly presented pictures. *Psychological Research, 43*(3), 277-292.

Balint, R. (1909). Seelenlähmung des Schauens, optische Ataxie, räumliche Störung der Aufmerksamkeit. *Monatsschrift für Psychiatrie und Neurologie, 25,* 51-81.

Biederman, I. (1972). Perceiving real-world scenes. *Science, 177,* 77-80.

Biederman, I. (Ed.) (1981). *On the semantics of a glance at a scene.* Hillsdale, NJ: Erlbaum.

Biederman, I., Rabinowitz, J. C., Glass, A. L., & Stacy, E. W. (1974). On the information extracted from a glance at a scene. *Journal of Experimental Psychology, 103*(3), 597-600.

Brandt, H. F. (1945). *The Psychology of Seeing.* New York: The Philosophical Library.

Brefczynski, J. A., & DeYoe, E. A. (1999). A physiological correlate of the 'spotlight' of visual attention. *Nature Neuroscience, 2*(4), 370 - 374.

Buswell, G. T. (1935). *How people look at pictures. A study of the psychology of perception in art.* Chicago: University of Chicago Press.

Carpenter, R. H. S. (1988). *Movements of the eyes.* London: Pion.

Cavegn, D. (1994). *Visual attention in fixation and saccade control: an analysis of express saccades.* Katholieke Universiteit te Leuven, Leuven.

Curcio, C. A., Sloan, K. R., Kalina, R. E., & Hendrickson, A. E. (1990). Human photoreceptor topography. *Journal of Comparative Neurology, 292,* 497-523.

Deubel, H., & Schneider, W. X. (1996). Saccade target selection and object recognition: evidence for a common attentional mechanism. *Vision Research, 36*(12), 1827-1837.

Enoch, J. M. (1959). Effect of the Size of a Complex Display upon Visual Search. *Journal of the Optical Society of America, 49,* 280-286.

Findlay, J. M., & Gilchrist, I. D. (1998). Eye Guidance and Visual Search. In G. Underwood (Ed.), *Eye Guidance in Reading and Scene Perception* (pp. 295-312). Oxford: Elsevier Science Ltd.

Findlay, J. M., & Gilchrist, I. D. (2003). *Active Vision.* New York: Oxford University Press.

Fischer, B. (1998). Attention in saccades. In R. D. Wright (Ed.), *Visual Attention* (Vol. 8, pp. 289-305). New York: Oxford University Press.

References

Fischer, B. (1999). *Blick-Punkte: Neurobiologische Prinzipien des Sehens und der Blicksteuerung.* Bern: Hans Huber.

Flagg, B. N. (1978). Children and television: effects of stimulus repetition on eye activity. In J. W. Senders, D. F. Fisher & R. A. Monty (Eds.), *Eye movements and the higher visual functions* (pp. 279-289). Hillsdale, NJ: Erlbaum.

Friedman, A., & Liebelt, L. S. (1976). On the time course of viewing pictures with a view towards remembering. In R. A. Monty & J. W. Senders (Eds.), *Eye Movements and Psychological Processes.* Hillsdale NJ: Lawrence Erlbaum Associates.

Furst, C. J. (1971). Automatizing of visual attention. *Perception & Psychophysics, 10*(2), 65-70.

Gilchrist, I. D., Brown, V., & Findlay, J. M. (1997). Saccades without eye movements. *Nature, 390,* 130-131.

Goldstein, E. B. (1999). *Sensation & Perception* (5 ed.). Pacific Grove CA: Brooks/Cole Publishing Company.

Gould, J. D. (1976). Looking at pictures. In R. A. Monty & J. W. Senders (Eds.), *Eye Movements and Psychological Processes* (pp. 307-321). Hillsdale NJ: Lawrence Erlbaum Associates.

Groner, R., Walder, F., & Groner, M. (1984). Looking at faces: Local and global aspects of scanpaths. In A. G. Gale & F. Johnson (Eds.), *Theoretical and Applied Aspects of Eye Movement Research* (22 ed., pp. 523-533). Amsterdam: Elsevier.

Henderson, J. M. (1992a). Object identification in context: The visual processing of natural scenes. *Canadian Journal of Psychology: Special Issue on Object and Scene Processing, 46,* 319-342.

Henderson, J. M. (1992b). Visual attention and eye movement control during reading and picture viewing. In K. Rayner (Ed.), *Eye Movements and Visual Cognition* (pp. 261-283). Berlin: Springer.

Henderson, J. M. (1993). Visual attention and saccadic eye movements. In G. d'Ydevalle & J. Van Rensbergen (Eds.), *Perception and Cognition* (pp. 37-50). Amsterdam: Elsevier.

Henderson, J. M. (2003). Human gaze control during real-world scene perception. *Trends in Cognitive Science, 7*(11), 498-504.

Henderson, J. M., & Ferreira, F. (2004). Introduction to the interface of vision, language, and action. In F. Ferreira & J. M. Henderson (Eds.), *The interface of language, vision, and action: Eye movements and the visual world.* New York: Psychology Press.

Henderson, J. M., & Hollingworth, A. (1998). Eye movements during scene viewing: an overview. In G. Underwood (Ed.), *Eye Guidance in Reading and Scene Perception* (pp. 269-293). Oxford: Elsevier Science Ltd.

Henderson, J. M., Pollatsek, A., & Rayner, K. (1989). Covert visual attention and extrafoveal information use during object identification. *Perception & Psychophysics, 45*(3), 196-208.

Henderson, J. M., Weeks, P. A. J., & Hollingworth, A. (1999). The effects of semantic consistency on eye movements during complex scene viewing. *Journal of Experimental Psychology: Human Perception and Performance, 25*(1), 210-228.

Henderson, J. M., Williams, C. C., Castelhano, M. S., & Falk, R. J. (2002). Eye movements and picture processing during recognition. *Perception & Psychophysics*.

Hodgson, T. L., & Müller, H. J. (1995). Evidence relating to premotor theories of visuospatial attention. In J. M. Findlay, R. W. Kentridge & R. Walker (Eds.), *Eye Movement Research* (pp. 305-316). Amsterdam: Elsevier.

Hoffman, J. E., & Subramaniam, B. (1995). The role of visual attention in saccadic eye movements. *Perception & Psychophysics, 57*(6), 787-795.

Irwin, D. E., Colcombe, A. M., Kramer, A. F., & Hahn, S. (2000). Attentional and oculomotor capture by onset, luminance and color singletons. *Vision Research, 40*, 1443-1458.

Ishihara, S. (1999). *Ishihara's Tests for Colour Deficiency. 38 Plates Edition*. Tokyo: Kanehara & Co., Ltd.

Itti, L., & Koch, C. (2000). A saliency-based search mechanism for overt and covert shifts of visual attention. *Vision research, 40*(10-12), 1489-1506.

Itti, L., & Koch, C. (2001). Computational modelling of visual attention. *Nature Reviews Neuroscience, 2*, 1-10.

Jost, T., Ouerhani, N., von Wartburg, R., Müri, R., & Hügli, H. (2005). Assessing the contribution of color in visual attention. *Computer Vision and Image Understanding Journal, 100*(1-2), 107-123.

Koch, C., & Ullman, S. (1985). Shifts in selective visual attention: towards the underlying neural circuitry. *Human Neurobiology, 4*, 219-227.

LaBerge, D., & Brown, V. (1989). Theory of attentional operations in shape identification. *Psychological Review, 96*(1), 101-124.

LaBerge, D., Carlson, R. L., Williams, J. K., & Bunney, B. G. (1997). Shifting attention in visual space: Tests of moving-spotlight models versus an activity-distribution model. *Journal of Experimental Psychology: Human Perception and Performance, 23*(5), 1380-1392.

Land, M. F., & Hayhoe, M. (2001). In what ways do eye movements contribute to everyday activities? *Vision Research, 41*, 3559-3565.

Land, M. F., Mennie, N., & Rusted, J. (1999). The roles of vision and eye movements in the control of activities of daily living. *Perception, 28*, 1311-1328.

References

Lindman, H. R. (1974). *Analysis of variance in complex experimental designs*. San Francisco, CA: W. H. Freeman.

Liversedge, S. P., & Findlay, J. M. (2000). Saccadic eye movements and cognition. *Trends in Cognitive Sciences, 4*(1), 6-14.

Loftus, G. R. (1981). Tachistoscopic simulations of eye fixations on pictures. *Journal of Experimental Psychology: Human Learning and Memory, 7*(5), 369-376.

Loftus, G. R. (1985). Picture perception: effects of luminance on available information and information-extraction rate. *Journal of Experimental Psychology: General, 114*(3), 342-356.

Loftus, G. R., & Mackworth, N. H. (1978). Cognitive determinants of fixation location during picture viewing. *Journal of Experimental Psychology: Human Perception and Performance, 4*(4), 565-572.

Mackworth, N. H., & Morandi, A. J. (1967). The gaze selects informative details within pictures. *Perception & Psychophysics, 2*, 547-552.

Maioli, C., Benaglio, I., Siri, S., Sosta, K., & Cappa, S. (2001). The integration of parallel and serial processing mechanisms in visual search: evidence from eye movement recording. *European Journal of Neuroscience, 13*, 364-372.

Mannan, S., Ruddock, K. H., & Wooding, D. S. (1995). Automatic control of saccadic eye movements made in visual inspection of briefly presented 2-D images. *Spatial Vision, 9*(3), 363-386.

Mannan, S., Ruddock, K. H., & Wooding, D. S. (1996). The relationship between the locations of spatial features and those of fixations made during visual examination of briefly presented images. *Spatial Vision, 10*(3), 165-188.

Mannan, S., Ruddock, K. H., & Wooding, D. S. (1997a). Fixation patterns made during brief examination of two-dimensional images. *Perception, 26*, 1059-1072.

Mannan, S., Ruddock, K. H., & Wooding, D. S. (1997b). Fixation sequences made during visual examination of briefly presented 2-D images. *Spatial Vision, 11*(2), 157-178.

Matin, E. (1974). Saccadic suppression: a review and an analysis. *Psychological Bulletin, 81*, 899-917.

McCullagh, P., & Nedler, J. A. (1989). *Generalized Linear Models* (2 ed.). London: Chapman & Hall.

Nelson, W. W., & Loftus, G. R. (1980). The functional field of view during picture viewing. *Journal of Experimental Psychology: Human Learning and Memory, 6*(4), 391-399.

Niebur, E., Itti, L., & Koch, C. (2001). Controlling the focus of visual selective attention. In L. Van Hemmen, D. E. & J. Cowan (Eds.), *Models of Neural Networks IV*. New York: Springer.

Noton, D., & Stark, L. (1971a). Scanpaths in eye movements during pattern perception. *Science, 171*, 308-311.

Noton, D., & Stark, L. (1971b). Scanpaths in saccadic eye movements while viewing and recognizing patterns. *Vision research, 11*, 929-942.

Ouerhani, N., von Wartburg, R., Hügli, H., & Müri, R. (2004). Empirical validation of the saliency-based model of visual attention. *Electronic Letters on Computer Vision and Image Analysis, 3*(1), 13-24.

Parkhurst, D., Law, K., & Niebur, E. (2002). Modeling the role of salience in the allocation of overt visual attention. *Vision Research, 42*(1), 107-123.

Parkhurst, D., & Niebur, E. (2003). Scene content selected by active vision. *Spatial Vision, 16*(2), 125-154.

Pierrot-Deseilligny, C., Ploner, C. J., Müri, R. M., Gaymard, B., & Rivaud-Péchoux, S. (2002). Effects of cortical lesions on saccadic eye movements in humans. *Annals of the New York Academy of Sciences, 956*, 216-229.

Pieters, R., Rosbergen, E., & Wedel, M. (1999). Visual attention to repeated print advertising: a test of scanpath theory. *Journal of Marketing Research, 36*, 424-438.

Posner, M. I. (1980). Orienting of attention. *Quarterly Journal of Experimental Psychology, 32*, 3-25.

Pratt, J., & Quilty, L. (2002). Examining the activity-distribution model of visual attention with exogenous cues and targets. *Quarterly Journal of Experimental Psychology, 55A*(2), 627-641.

Rayner, K. (1984). Visual selection in reading, picture perception, and visual search. A tutorial review. In H. Bouma & D. G. Bouwhuis (Eds.), *Attention and Performance X. Control of Language Processes*. London: Erlbaum.

Rayner, K., & Pollatsek, A. (1992). Eye movements and scene perception. *Canadian Journal of Psychology, 46*(3), 342-376.

Rizzolatti, G., Riggio, L., Dascola, I., & Umiltá, C. (1987). Reorienting attention across the horizontal and vertical meridians: Evidence in favor of a premotor theory of attention. *Neuropsychologia, 25*, 31-40.

Ryan, J. D., Althoff, R. R., Whitlow, S., & Cohen, N. J. (2000). Amnesia is a deficit in relational memory. *Psychological Science, 11*(6), 454-461.

Saida, S., & Ikeda, M. (1979). Useful visual field size for pattern perception. *Perception & Psychophysics, 25*(2), 119-125.

Salthouse, T. A., & Ellis, C. L. (1980). Determinants of eye-fixation duration. *American Journal of Psychology, 93*(2), 207-234.

Sanders, A. F., & Donk, M. (1996). Visual search. In A. F. Sanders & O. Neumann (Eds.), *Handbook of Perception and Action: Attention III* (pp. 43-77). London: Academic Press.

Saslow, M. G. (1967). Latency for saccadic eye movements. *Journal of the Optical Society of America, 57*, 1030-1033.

References

Schiller, P. H. (1998). The neural control of visually guided eye movements. In J. E. Richards (Ed.), *Cognitive Neuroscience of Attention: A Developmental Perspective*. Hillsdale, NJ: Lawrence Erlbaum Associates.

Schneider, W. X., & Deubel, H. (1995). Visual attention and saccadic eye movements: Evidence for obligatory and selective spatial coupling. In J. M. Findlay, R. W. Kentridge & R. Walker (Eds.), *Eye Movement Research* (pp. 317-324). Amsterdam: Elsevier.

Schyns, P. G., & Oliva, A. (1994). From blobs to boundary edges. *Psychological Science, 5*(4), 195-200.

Shioiri, S., & Ikeda, M. (1989). Useful resolution for picture perception as a function of eccentricity. *Perception, 18*(3), 347-361.

Stark, L. W., & Choi, Y. S. (1996). Experimental metaphysics: the scanpath as an epistemological mechanism. In W. H. Zangemeister, H. S. Stiehl & C. Freksa (Eds.), *Visual Attention and Cognition* (pp. 3-69): Elsevier.

Stevens, A. L., & Rumelhart, D. E. (1975). Errors in reading: Analysis using an augmented transition network model of grammar. In D. A. Norman & D. E. Rumelhart (Eds.), *Explorations in Cognition*. San Francisco, CA: Freeman.

Tatler, B. W., Gilchrist, I. D., & Land, M. F. (2005). Visual memory for objects in natural scenes: from fixations to object files. *Quarterly Journal of Experimental Psychology, 58A*(5), 931-960.

Treisman, A. (1983). The role of attention in object perception. In O. J. Braddick & A. C. Sleigh (Eds.), *Physical and biological processing of images*. Berlin: Springer.

Treisman, A., & Gelade, G. (1980). A feature-integration theory of attention. *Cognitive Psychology, 12*, 97-136.

van Diepen, P. M. J., DeGraef, P., & d'Ydevalle, G. (1995). Chronometry of foveal information extraction during scene perception. In J. M. Findlay, R. W. Kentridge & R. Walker (Eds.), *Eye Movement Research* (pp. 349-362). Amsterdam: Elsevier.

von Wartburg, R., Ouerhani, N., Pflugshaupt, T., Nyffeler, T., Wurtz, P., Hügli, H., et al. (2005). The influence of colour on oculomotor behaviour during image perception. *NeuroReport, 16*(14), 1557-1560.

Wandell, B. A. (1995). *Foundations of Vision*. Sunderland, MA: Sinauer Associates, Inc.

Wright, R. D. (1998). *The control of visual attention* (Vol. 8). New York: Oxford University Press.

Wright, R. D., & Ward, L. M. (1998). The control of visual attention. In R. D. Wright (Ed.), *Visual Attention* (Vol. 8, pp. 132-186). New York: Oxford University Press.

Yarbus, A. L. (1967). *Eye Movements and Vision*. New York: Plenum Press.

Appendix A: Images used in Experiment 1

Category "Traffic"

Category "Village"

Category "Waters"

Appendix B: Result tables for Experiment 1

Table B-1: Histogram analysis of saccade amplitude distribution

Saccade amplitude (bin centre)	Size 1 Frequency	Size 1 Absolute %	Size 2 Frequency	Size 2 Absolute %	Size 3 Frequency	Size 3 Absolute %	Size 4 Frequency	Size 4 Absolute %
0	596	2.01%	500	1.45%	358	.97%	360	.93%
0.5	2077	14.03%	1494	8.65%	1228	6.65%	1071	5.53%
1	2385	16.11%	1755	10.16%	1377	7.46%	1212	6.26%
1.5	2181	14.73%	1535	8.88%	1278	6.92%	1120	5.79%
2	1857	12.54%	1528	8.84%	1179	6.38%	1084	5.60%
2.5	1545	10.43%	1315	7.61%	1084	5.87%	1023	5.28%
3	1120	7.56%	1157	6.70%	1015	5.50%	887	4.58%
3.5	985	6.65%	1022	5.91%	901	4.88%	832	4.30%
4	705	4.76%	907	5.25%	914	4.95%	753	3.89%
4.5	523	3.53%	865	5.01%	840	4.55%	782	4.04%
5	386	2.61%	802	4.64%	777	4.21%	808	4.17%
5.5	278	1.88%	716	4.14%	789	4.27%	717	3.70%
6	176	1.19%	636	3.68%	752	4.07%	681	3.52%
6.5	98	.66%	569	3.29%	692	3.75%	662	3.42%
7	85	.57%	505	2.92%	625	3.38%	655	3.38%
7.5	47	.32%	471	2.73%	616	3.34%	599	3.09%
8	29	.20%	383	2.22%	548	2.97%	548	2.83%
8.5	16	.11%	298	1.72%	511	2.77%	561	2.90%
9	6	.04%	238	1.38%	420	2.27%	552	2.85%
9.5	5	.03%	215	1.24%	382	2.07%	456	2.36%
10	2	.01%	158	.91%	304	1.65%	440	2.27%
10.5	2	.01%	85	.49%	303	1.64%	411	2.12%
11	0	.00%	115	.67%	269	1.46%	348	1.80%
11.5	0	.00%	86	.50%	244	1.32%	317	1.64%
12	0	.00%	48	.28%	187	1.01%	334	1.73%
12.5	1	.01%	40	.23%	171	.93%	250	1.29%
13	0	.00%	21	.12%	151	.82%	235	1.21%
13.5	0	.00%	31	.18%	122	.66%	206	1.06%
14	0	.00%	17	.10%	105	.57%	191	.99%
14.5	0	.00%	10	.06%	94	.51%	157	.81%
15	0	.00%	3	.02%	80	.43%	138	.71%
15.5	0	.00%	3	.02%	60	.32%	116	.60%
16	0	.00%	3	.02%	57	.31%	108	.56%
16.5	1	.01%	0	.00%	43	.23%	111	.57%
17	0	.00%	0	.00%	33	.18%	105	.54%
17.5	0	.00%	0	.00%	24	.13%	106	.55%
18	0	.00%	0	.00%	26	.14%	85	.44%
18.5	0	.00%	0	.00%	16	.09%	64	.33%
19	0	.00%	0	.00%	10	.05%	65	.34%
19.5	0	.00%	0	.00%	14	.08%	56	.29%
20	0	.00%	0	.00%	11	.06%	48	.25%
20.5	0	.00%	0	.00%	12	.06%	41	.21%
21	0	.00%	0	.00%	13	.07%	35	.18%
21.5	0	.00%	0	.00%	3	.02%	40	.21%
22	0	.00%	0	.00%	1	.01%	27	.14%
22.5	0	.00%	0	.00%	2	.01%	14	.07%
23	0	.00%	0	.00%	3	.02%	19	.10%
23.5	0	.00%	0	.00%	0	.00%	16	.08%
24	0	.00%	0	.00%	1	.01%	22	.11%

Table B-2: Saccade amplitude distribution over all four image sizes, reparsed with different eye movement parsing parameters. The values represent the number of saccades per bin.

Saccade amplitude (bin centre)	Parsing parameters: speed/acceleration thresholds			
	15/1800	18/2600	22/3800	35/9500
0.1	3222	2138	1575	471
0.2	2212	1770	1419	1161
0.3	1926	1570	1191	1148
0.4	1832	1636	1368	1642
0.5	1757	1681	1514	1530
0.6	1829	1737	1630	1373
0.7	1670	1668	1589	1509
0.8	1694	1737	1697	1544
0.9	1613	1644	1572	1689
1.0	1651	1644	1581	1519
1.1	1555	1574	1665	1474
1.2	1615	1660	1618	1432
1.3	1560	1506	1548	1452
1.4	1488	1557	1509	1466
1.5	1468	1495	1518	1384
1.6	1417	1423	1354	1368
1.7	1336	1358	1376	1355
1.8	1306	1362	1345	1295
1.9	1331	1288	1272	1296

Table B-3: Temporal course of saccade amplitude change. The values represent the temporal variation of amplitudes in degrees visual angle, irrespective of the overall mean values. The main effect values represent the means over all four image sizes.

	temporal window				
	1st	2nd	3rd	4th	5th
Size1	-0.157	0.138	0.140	-0.037	-0.076
Size2	-0.076	0.200	0.045	-0.015	-0.132
Size3	-0.053	0.175	0.138	-0.090	-0.205
Size4	0.118	0.360	-0.033	0.001	-0.381
Main effect	-0.042	0.218	0.073	-0.035	-0.199

Appendix C: Images used in Experiment 2

— Appendices

Appendix D: Result tables for Experiment 2

Table D-1: Results for temporal course of fixation duration analysis. Values are given in ms.

viewing task	viewing	time window									
		1	2	3	4	5	6	7	8	9	10
free viewing	1st	272.3	300.7	294.4	297.6	314.3	308.2	320.7	311.8	330.5	325.1
	2nd	293.7	311.2	332.5	332.7	348.5	341.0	337.6	367.4	336.9	338.3
recognition	1st	301.6	320.3	325.0	317.6	336.4	336.0	352.1	353.8	362.5	350.1
	2nd	340.0	353.7	346.7	365.8	358.5	379.9	360.5	354.0	367.9	368.7
detail memory task	1st	277.8	300.9	305.1	303.1	300.4	304.4	305.7	300.4	305.9	307.6
	2nd	285.1	317.7	312.5	315.7	310.3	325.8	318.6	319.8	325.1	323.2

Table D-2: Results for temporal course of saccade amplitude analysis. Values are given in degrees.

viewing task	viewing	time window									
		1	2	3	4	5	6	7	8	9	10
free viewing	1st	4.850	5.541	5.643	5.707	5.517	5.388	5.440	5.619	5.291	5.262
	2nd	4.556	4.953	5.204	4.962	5.192	5.027	5.000	4.761	4.745	4.550
recognition	1st	4.827	5.480	5.618	5.522	5.693	5.476	5.503	5.150	5.563	5.364
	2nd	4.429	5.183	5.129	5.219	5.359	5.107	5.287	5.033	5.072	4.907
detail memory task	1st	4.693	5.408	5.385	5.371	5.381	5.300	5.269	5.630	5.453	5.329
	2nd	4.397	5.040	5.182	5.078	5.129	5.092	5.086	5.340	4.993	5.002

Table D-3: Scanpath analysis results: Participant number (pnb), instruction (instr; 1 = free, viewing, 2 = recognition, 3 = detail memory task), number of images with valid data (nbimgs), raw number of scanpaths for n-fixation sequences (nb$n$), total number of scanpaths (total), and number of scanpaths per image (paths/img).

pnb	instr	nbimgs	nb2	nb2	nb4	nb5	nb6	total	paths/img
942	1	48	36	3	2	0	0	41	0.854
948	2	47	36	8	1	0	0	45	0.957
941	1	48	30	6	0	1	0	37	0.771
946	1	46	35	7	0	0	0	42	0.913
958	2	45	41	8	2	0	0	51	1.133
943	1	48	44	9	2	0	0	55	1.146
971	3	48	55	11	2	0	1	69	1.438
959	2	46	31	6	0	0	0	37	0.804
966	1	48	27	2	2	0	0	31	0.646
955	2	48	46	8	2	0	1	57	1.188
962	2	44	32	7	1	0	0	40	0.909
979	3	47	41	4	2	0	0	47	1.000
954	1	48	43	4	0	0	0	47	0.979
956	1	48	39	9	1	0	0	49	1.021
981	3	38	20	3	0	0	0	23	0.605
985	3	48	67	4	2	0	0	73	1.521
945	1	32	16	4	0	0	0	20	0.625
961	2	48	56	10	3	1	1	71	1.479
949	2	48	52	4	0	0	0	56	1.167
970	3	48	68	11	1	1	0	81	1.688
965	3	47	72	5	2	0	0	79	1.681
944	1	48	27	5	0	0	0	32	0.667
983	3	48	24	5	0	0	0	29	0.604
969	1	48	44	4	0	1	0	49	1.021
980	3	48	49	9	0	1	0	59	1.229
953	1	48	33	6	3	0	0	42	0.875
976	3	47	35	3	0	1	0	39	0.830
960	2	47	32	5	1	0	0	38	0.809
978	3	48	46	3	1	0	0	50	1.042
975	3	46	39	7	3	1	0	50	1.087
940	1	48	40	8	0	0	0	48	1.000
957	2	46	44	7	1	0	0	52	1.130
989	3	48	61	11	3	1	0	76	1.583
939	1	48	34	7	1	1	0	43	0.896
947	2	40	25	6	0	1	0	32	0.800
964	2	48	18	1	0	0	0	19	0.396
1002	2	48	31	6	0	1	0	38	0.792
1012	3	48	47	7	1	0	0	55	1.146
1011	3	48	39	3	0	0	0	42	0.875
1006	3	44	31	4	0	0	0	35	0.795
1013	3	41	22	0	0	0	0	22	0.537
1005	2	48	31	5	0	0	0	36	0.750
1004	1	48	33	3	0	0	0	36	0.750
1015	3	48	35	5	5	0	0	45	0.938
1016	1	47	35	3	0	0	1	39	0.830
1026	2	47	42	7	0	0	0	49	1.043
1024	2	46	48	7	2	0	0	57	1.239
1025	2	48	53	8	0	0	0	61	1.271
1050	1	47	39	8	1	0	0	48	1.021
1079	2	48	66	11	1	2	0	80	1.667
1084	2	48	20	3	1	0	0	24	0.500
1082	1	44	39	8	1	1	0	49	1.114
1086	3	48	39	13	1	3	0	56	1.167
1088	1	44	14	0	0	0	0	14	0.318

VDM Verlagsservicegesellschaft mbH

Die VDM Verlagsservicegesellschaft sucht für wissenschaftliche Verlage abgeschlossene und herausragende

Dissertationen, Habilitationen, Diplomarbeiten, Master Theses, Magisterarbeiten usw.

für die kostenlose Publikation als Fachbuch.

Sie verfügen über eine Arbeit, die hohen inhaltlichen und formalen Ansprüchen genügt, und haben Interesse an einer honorarvergüteten Publikation?

Dann senden Sie bitte erste Informationen über sich und Ihre Arbeit per Email an *info@vdm-vsg.de*.

Sie erhalten kurzfristig unser Feedback!

VDM Verlagsservicegesellschaft mbH
Dudweiler Landstr. 99
D - 66123 Saarbrücken

Telefon +49 681 3720 174
Fax +49 681 3720 1749

www.vdm-vsg.de

Die VDM Verlagsservicegesellschaft mbH vertritt

Printed by Books on Demand GmbH, Norderstedt / Germany